THINGS THAT ARE

Amy Leach grew up in Texas. She holds an MFA in creative nonfiction from the University of Iowa, and her work has appeared in numerous literary journals and reviews. She has been recognised with the Whiting Writers' Award and a Rona Jaffe Foundation Award. She lives in Montana.

Charlotte

Dad
♡

THINGS THAT ARE

ENCOUNTERS WITH PLANTS, STARS AND ANIMALS

AMY LEACH

CANONGATE

This paperback edition published in 2017 by Canongate Books

First published in Great Britain in 2013 by Canongate Books Ltd, 14 High Street, Edinburgh EH1 1TE

canongate.co.uk

2

First published in the United States of America in 2012 by Milkweed Editions, 1011 Washington Avenue South, Suite 300, Minneapolis, Minnesota 55415

British Library Cataloguing-in-Publication Data
A catalogue record for this book is available on
request from the British Library

ISBN 978 1 78689 355 0

Interior design by Gretchen Achilles/Wavetrap Design
Typset in Mrs Eaves

Printed and bound in Great Britain by Clays Ltd, St Ives plc.

TO MATTHEW

Contents

All things that are, are equally removed from being nothing.

—JOHN DONNE

I was extremely gratified to think, that if I had pleasures they knew nothing of, they had also some into which I could not possibly enter.

—JOHN KEATS

writing about country dancers

Donkey Derby

Usually all we have to do when we go a-conquering is build a boat, find a benefactress, recruit a ribald crew, and wear radiant glinting helmets. With these four easy steps my kind has conquered faraway lands, and seas and moons and molecules. However, even after thousands of years, we have had no luck conquering Tomorrow. Over and over again, we have set sail in pursuit of Tomorrow only to discover Tomorrow's antecedents. It is a recurring disappointment, like never leaving Spain.

Perhaps with some things, the only way to conquer them is to abolish them. If only the Earth didn't turn there would be no Tomorrow to confound us, there would just be Today and Tonight, and they would hold still like Peru; they would be clearly marked on the atlas. We could build colonies in each of them and travel back and forth at will; days and nights wouldn't come upon us, we would come upon them.

How do you stop a thing from spinning? To stop a prayer wheel from spinning you take it away from the practitioner. Once he is no longer rhythmically twitching his wrist it will slow down and stop. But if the Earth is a prayer wheel it's a prayer wheel we're glued to like miniature symbols. What can we do but yell at the practitioner to please stop spinning it and generating all of those chancy tomorrows.

But while we have yelling moods, and imperial moods, we also have guessing moods. Even with our fine record of conquests, there remain a few things we do not have atlases for, like tomorrow and the rain and the gods and donkeys. This is the sweet stuff of gambling – the chancier the better. Betting on wild donkeys at the Kentucky Derby is even more fun than betting on thoroughbreds: with wild donkeys from the salt flats there are no tired conventions like 'early moderate tempo' or 'tactical speed' or even 'forward progression'. A donkey derby is nothing but upsets, from start to finish. And betting on the gods is better still; it feels like placing bets on Thelonious Monk's ten fingers – which finger will play what key next!

Tonight, on this the latest antecedent to tomorrow, it is starry out and I am not in a conquering mood. Come and miss the boat with me. Come and play some guessing games. We'll read aloud the illegible electric green script of the northern lights; we'll speculate about which star in the next ten thousand years is going to go supernova. Then we'll listen to a recording of 'Epistrophy'. I'll wager on his left thumb, you take whichever finger you want, and with the mad currency we collect from each other I'll buy you rain, you buy me snow, and we'll go in together for sunshine for the grass and the clover and the delicious prickly thistles.

I.

THINGS

OF

EARTH

Trappists

I am a Trappist like the trees,' the lily thought to herself as she let the breeze move her but said no words to it. 'I am a Trappist like the lily,' the creek thought to himself as he swelled with pearly orange fishes but declined to converse with them. 'We are Trappists like the creek,' thought the raindrops, as they filled the pond with fresh cloud water, or mixed with the juice of a fallen cherry, or came to rest deep in the dirt, and everywhere neglected to introduce themselves. 'I am a Trappist like the rain,' thought the tree, as she felt the taciturn rain dripping off her warm needles onto the ground and the wet birds returning, and she made no speeches. 'I am a Trappist like the trees,' the Trappist thought to himself as he walked into the forest, as he let the lily, the creek, and the fishes and the rain move him, and he said nothing.

In Which the River Makes Off with Three Stationary Characters

In the seventeenth century, his Holiness the Pope adjudged beavers to be fish. In retrospect, that was a zoologically illogical decision; but beavers were not miffed at being changed into fish. They decided not to truckle to their new specification, not to be perfect fish, textbook fish; instead they became fanciful fish, the first to have furry babies, the first to breathe air and the first fish to build for themselves commodious conical fortresses in the water. If Prince Maximilian, travelling up the Missouri River, had taken it in mind to recategorise them as Druids or flamingos, beavers would have become toothy Druids, or portly brown industrious flamingos.

The beavers' reaction to the papal renaming highlights two of their especial qualities: their affability and their unyieldingness. They affably yield not. They live in cold wet water but are warm and dry in their oily parkas. If they are deemed fishes, they respond by becoming lumberjacking fishes. They-of-the-Incisors are puppets of no pope, and puppets of no river, either. The river, where the beaver lives, is at cross-purposes with the beaver, in that it is tumbling away, while the beaver wants to produce kindred at One Address. An animal more contrary than the beaver would build a grumpy shanty of sticks in the forest; an animal less contrary the river would drag and distract and make into memorabilia.

The Moon also graces the water without getting floated off its feet, but effortlessly, while beavers have to work as hard as derricks. What it takes for them to prepare a mansion for themselves, in the midst of gallivanting water, with nothing to wield but short arms and long teeth, is constant botheration; they chew and lug and wrestle logs all night long, unless wolverines or humans visit. When these disputatious creatures turn up the beavers swim to the underwater tunnel to their cabin and climb up and hide out, for they do not like to fight.

Beaver babies cannot sink or swim when they are born; if they accidentally slip down the tunnel into the water they are like tiny complaining pontoons. In several hours, though, they can swim, front paws up by their chins, paddling with their huge ducky backfeet; and by May, after drinking a month's worth of fat buttery milk, the burnished brown babies are working, swimming their little twigs to the dam to help with repairs.

They will never stop working thereafter, unless one of them happens to be voted an extraneous beaver, during the periodic population control that beaverocracies exercise. Even the most agreeable animals can only stand so many of themselves per pond. An expelled beaver by himself will just crouch in a mudhole, like a mouldywarp, and have time to get lost in thought; unlike his cousins and brothers and grandmothers chewing down four hundred trees every year; careening away when the trees start to fall over; shuffling back to drag the timber through the grass, wrangling poplars and birches and piano benches – whatever is wooden; digging log flumes and making log-rolling paths, swimming the trees down the stream, shoving them together into a dam, making the dam wider each night and higher and higher, repairing the dam when a leak springs; heaping up a house of aspens, trundling down the shore with armfuls of muddy rubble to plaster the walls with, repairing the roof after a bear performs roof meddlement, plunging cherry trees underwater, in order to have sumptuous foodstuffs in the larder in January, for the Feast of the Bean-King, when ponds are covered with two feet of ice.

With their powers of reorganisation, beavers recapitulate the creation of the world, gathering water together in one place and making dry mounds appear in another. In fact they were probably there at the original one, acting as auxiliary spirits, helping to impose landscape on the mishmash, heaping up dry land for the earthgoers and corralling the waters for the swimming animals. How boggy and spongelike would the world be without beavers to divide it up! What type of tenants would we attract but bladderworts and mudpuppies!

But even if they were the ones who installed it, beavers are still subject to topography. A river's patron-glacier may melt so catastrophically that the river overthrows a beaver dam, and before they can mobilise Barrier Repair the beaver colony gets bundled off to sea, like fat astonished fishes. Though octopuses make sense in the ocean, beavers and cactuses and pencilmakers do not. When they get there the ocean must derange them, making them delirious, because the sound of water is what triggers their gnawing reflex. As soon as they hear the burbly gushing of a stream, beavers speed to the nearest trees to chisel girdles around their trunks so they go whomping down and then they can stuff them into the chatterboxy river to strangulate it into silence. But the ocean is a wilderness of chatter, and not in all the forests of the world are there enough trees to muzzle its splashing, sloshing, gurgling, yammering, yackety-yacking waves.

LATE IN LIFE, when salmon are old salts, long having lived at sea, they decide to hoist themselves up a river, back to the scene of their nativity, with its particular mushroom-and-lily perfume. They smell their way there. If you subtracted the mushrooms and the lilies and substituted some frumenty and glögg and sagittaries with beer-breath, how would the fishes recognise their birthplace? They would slog right past it, up a tributary creek until they got to the icy seep of the river's tiny origination faucet.

The brides and grooms toil up their nine-hundred-mile aisle for weeks and weeks to reach the mushroomy altar. Once there, they deposit their ingredients into the bottom-gravel — ingredients which when congealed will result in seven thousand black-eyed eggs. When these spiffy little fishes have hatched from their eggs and self-excavated from the gravel, they hide in crannies and absorb the yolks bequeathed to them. Then the bequests run out; then they swim in place and hold their mouths open to swallow the crustaceans drifting by. Not inheriting little anchors to hook into the riverbed, the fishes countervail the flow of the river by plying their fins, making endless varieties of strokes, all of which mean No. Maybe it feels like maintaining the same longitude on a steam train going east. Maybe it feels like being tossed endless apples while trying to retain a total of zero apples.

The little open-mouthed fishes swim against the river for one-and-a-half years, returning to where they are every moment, exercising all their hydrodynamical competence in not being spun around to the left or to the right, in not pitching head-over-tail or tail-over-head, in not getting rolled sideways like cartwheels in the current — the influential current of ambient custom which would draw all creatures pitching, yawing, rolling down its sinuous swallowing throat, all creatures become gobbets.

The salmon fry live in this milieu as dissenters, like the beavers; and they also labour relentlessly to stay in one place — not by concocting deluxe accommodations for themselves, but by sculling their delicate translucent fins all winter, spring, summer, fall. Their wilfullness is their anchor. You would think, after so many months, that the anchor was permanent, that No was the only word they knew, that they would forever correct for the vector of the river.

Then something countervails their will to countervail. Their will tips over and they let the water swoop them away, spilling them backwards down churning frothy staircases of rocks, rushing them through ice-blue hourglasses between basaltic cliffs covered with maidenhair ferns, flicking them down to the fluted mud, where rest jettisoned peace pipes and scarfpins, streaming them under mossy sodden maples and sodden mossy yews, crisscross-fallen in the water, pouring them over shallow stony flats and dilly-dallying them around lazy crooks and switches, past yellow monkeyflowers on the shore and elfin groves of watercress, depositing them at last in the very vasty habitat of octopuses.

SOMETIMES ON A PORCH in June, a girl begins to plunk her banjo; and after a spell of stillness, while the sound travels down their ear crinkles into their inmost feeling-chambers, the music starts to dance the people passing by. They toss like puppets on a bouncing sheet; like boys without a boat; they swing like weeds in the wind; they leap heptangularly about, dancing eccentric saltarellos, discovering that their springs are not so rusty.

For even if you have built masterful aspen castles in your mind, have toppled whole forests to throttle the writhing elements into a liveably serene personal pond; if you have longtime sculled your ingenious fins to withstand the tumble-crazy currents; there is music that will dissolve your anchors, your sanctuaries, floating you off your feet, fetching you away with itself. And then you are a migrant, and then you are amuck; and then you are the music's toy, juggled into its furious torrents, jostled into its foamy jokes, assuming its sparklyblue or greenweedy or brownmuddy tinges, being driven down to the dirgy bottom where rumble-clacking stones are lit by waterlogged and melancholy sunlight, warping back up to the surface, along with yew leaves and alewives and frog bones and other strange acquisitions snagged and rendered willy-nilly by the current, straggling away on its rambling cadenzas, with ever-changing sights — freckled children on the bank, chicken choirs, brewing thunderclouds, june bugs perched in wild parsley — until it spills you into a place whose dimensions make nonsense of your heretofore extraordinary spatial intelligence.

Goats and Bygone Goats

It is too bad that sound waves decay. If they did not, we would still be able to hear melodies by Mesomedes, and Odo of Cluny playing his organistrum. We would hear extinct toxodons, and prehistoric horses wearing pottery bells, and dead bats chewing crackly flies. We could hear the goats of the past – the old English milch goats, the fatlings of Bashan, thirsty peacock goats, Finnish Landrace goats bleating for their kids, baby Göingegets grizzling for their mothers, and wild mountain ibexes protesting hoarsely at being made to live in the *Jardin des Plantes* in Paris. The world, full of past sound, would be like the sky, full of past light. The world would be like the mind, for which there is no once.

But the material – air – that makes sound possible also exhausts it. In an unresounding atmosphere, one is left to surmise about the sundered past; about, for example, the precursors of Hungarian improved goats. What were Hungarian unimproved goats like? Probably like any unimproved goats — lousy and ticky, with horehound, clover seeds, and other faults tangled in their fleece; prone to poking each other with their horns. You can make a goat easier to live with by confiscating his horns; but without them the whole goat is more likely to be confiscated. And most goats, except for fainting goats, are not meant for stealing.

A fainting goat often serves as special companion to a herd of sheep. When they get rattled at, screeched at, hollered at, fainting goats sprint away for a second and then freeze, toppling like upended chairs. This is not floppy kid syndrome, nor mad staggers — which entail blindness and spinal disintegration. Fainting goats just fall over for a few seconds, muscles rigidly locked, fully conscious, like terrified figurines. So when a coyote rushes from behind a boulder, the goat is stationary, available, and the cream puffs can totter away.

Experience is so capricious. Now one is supple, now changed to stone. Now the goat in the timothy meadow is standing on lithe legs, magic legs — legs ready to grant her pixiest wishes. Then she hears a growl, or a cracking twig, and her sensitive legs turn to iron pokers, and she tumbles and cannot move. The salty toads vault over her, the thistles nod, the withy-wind waves and the witch moths float by. O grey goat, what can your wishes do without your magic legs? It avails to be a sheep.

Sometimes it avails to be a goat. When the grass withers away in Morocco, sheep will stumble dully along, thinking horizontal thoughts. *No grass . . . no grass.* But goats look up, start climbing trees. Even with fifteen goats in its crooked knobbly arms, lunching on its suspended fruit, the argan tree is trusty, for its roots are deep clutched in the earth. It does not wait for mercy to fall from heaven: when the sky is dry the tree fathoms the earth, stabbing its roots down until they discover buried rain — rimstone pools undreamt of by the grass lying vegetably on the surface. The argan tree drinks, drinks, while the grass waits, waits. Grass waits for water like sheep wait for grass.

Of the fleecy species, goats are slightly more universal than sheep. In the sixteenth century, explorers sailing around the world took goats with them and sowed them on miscellaneous islands. Wisely, they did not disperse sheep, who would have made hapless pioneers, or a specialist — after even a week on a tropical island, pygmy rabbits, who eat only sagebrush, would have been hot, sandy, sunk in blurry starvation dreams.

But goats are generalists: the world is their meadow. Leave them on an island — they will not spend all their energy on refusal and regret but will experiment until they find something new to eat, life sufficient condiment for the scraggliest fare. Put them in a barn with frocks and cigars and political pamphlets and toy blocks and banjos and yo-yos and frog leather — they will try everything, even the barn studs. They investigate by chewing and chew more than they swallow, in contrast to sharks who investigate by swallowing and swallow more than they chew.

How terrible for the pioneer goats in the end, when the sailors returned! But how splendid in the interim — between the sowing and the reaping! After lives of being followed around by people with shears, people with milk pails, people with scalpels; sharing fields with sheep who never stop communicating; being corralled, prodded, nipped at and yapped at by border collies in the wroth winter weather; then after coopy months on a tossing greasy ship — to be lowered into a wherry and rowed to shore on a palmy blue evening and left behind, to rest their shipworn bodies on the quiet beach for the night, and open their eyes in the morning to lagoon light, translucent yellow fruit and turquoise bird wings and emerald dew-drippy leaves! To be free! On a ferny island! With sweet rainwater and fellow goats! O life like wine!

On some ferny islands the goats ran wild, became as successful as flames (fire is a generalist too). Pinta

Island, for example, was fernier before the goats landed and took a fancy to the tree ferns, which giant tortoises had always used as shady canopies. Since the goats ate the tortoises' food as well, Pinta tortoises eventually lost their grip. All except for Lonesome George. For thirty-five years Lonesome George lived by himself on Pinta Island until he was moved to an institution and beatified alive.

To prevent other tortoises on other islands from becoming similarly rare, similarly sainted, some people have proposed the importation of dingoes. The trouble is, after the dingoes finished the goats they might eat the natives, so crocodiles would have to be introduced to eat the dingoes. A succession of increasingly dangerous animals would have to be sailed to the island until someone would inevitably have to bring thirty hippopotamuses across the ocean and set them loose to squash everything, a stable but sad climax. To circumvent this and other onerous scenarios there is another solution: guns. Some tortoise advocates just shoot the goats from helicopters. If it seems like the noise would bother the tortoises, it does not — such innocents do not know a bang-bang from a ding-dong.

Goats appear to have more misgivings, though, for there are always hundreds who evade the helicopter sessions. To locate these fugitives, a goat will be trapped, given a radio collar and sent back into the bush to search them out, for goats do not like to be alone. If you only own one goat, instead of two, she will flock to you, scramble into your car, chew through the palings between you, climb the fire escape, walk along narrow carpentry, along a drainpipe, over a roof, and bleat a million bleats, just to be with you, to have you rub her long, heavy ears and stroke her withers. So Radio Goat, he goes looking for companions, and when he finds them his collar advises the helicopter, whereupon all the goats are disembodied.

Elsewhere, people are trying to reembody certain goats, like the bucardo, a Spanish mountain goat. (Mountain goats are as agile as tightrope dancers, but who can be agile on a landslide?) The last bucardo was found with her head crushed by a falling tree. It is an old truism that no bygone goat rises again, but this is no longer certain, because someone was careful to preserve a bucardo ear, which is more than anyone did for the quagga, and someday from this ear bucardos may laboratorily spring.

FOR NOW, bucardos dwell in an ear. They dwell in the potential world, where they are pushing their noses into soft moss and eating potential rosinweed and glory peas in the cold glittering sleet, growing thick brown wool, bearing triplets of wriggly kids whose twisty zigzags tempt potential people to laughter, sharing a mountain with Etruscan shrews and mouflon sheep and flaxen boarlets and fat battling marmots and napping rusty squirrels and rosy goldfinnies and hazel dormice, and potential otters paddling in streams threaded with slender blennies. Bucardos are caught in the potential world as the fatlings of Bashan are caught in the past world; as fainting goats are caught in the actual world. There is no ladder out of any world; each world is rimless.

They say if you are leading your flocks from a depleted field to a field of fresh Spanish broom, and you stop to rest, the sheep will stand there wondering what on earth is going on. But the goats lie down, switching nimbly from travelling to resting to leaping to ruminating; from barrelling into each other, horns first, to listening spellbound to the tipple flute; from munching on lantana and woody weeds to gathering together – as the sun sets on whichever implausible world they inhabit – to sink into reasonless, companionate sleep.

Talent

On baby wrens hatched in a hanging flower basket, the sun shines through silky leaves and flapping flowers and glittery rain, which inspires earthed worms to unearth themselves. The spring winds may send the nest of wrens swinging, but swinging with geraniums. In the warm flowered sunlight, the newborn wrens begin to riot. Their riots are perpetually rewarded with worms. The only hint of winter for the wren, when the wren is young, is that flowers close at night.

Perhaps if penguin eggs were hatched in geranium baskets, penguins would be better at rioting. As it is, when an emperor penguin hatches out of its egg, it presses its two short flippers against its body and stands bunched on its father's feet, on the ice, in the dark. It is born midwinter in Antarctica, where winter is unremittingly dark. The father stands over; but he is starving; he hasn't eaten in four months. Hundreds of miles out in the ocean, the mother has been winter fishing for squid and fish a thousand feet down in the dark sea for two months, and now she is hopefully walking back across a hundred miles of ice, and she is hopefully fat. But she doesn't walk very fast because her knees don't bend.

The reason penguins don't place their egg at an easier distance from the sea is that their continent has meltable edges. Sorry the infant who stands close to sea, for spring may send it drifting off on a detached floe of ice; a straying acre bearing away a stationary, pear-shaped bit of fluff. How do you locate a wave?

If the female doesn't come back soon after the chick hatches, the starving father will have to shuffle, or else snow-paddle on his stomach, away to the sea. The saggy hatchling will stand on the ice, as far from fish as from flowers. Another mother returning to the colony with food will not help; if she finds herself being torn after by three frantically hungry chicks, she will out-tear two of them until she is pursued only by her true chick. So true chicks will be plumper than spurious chicks. Spurious chicks will close their black eyes and bunch down on the bare ice, unbabied, in the darkest, stabbingest winter on earth.

But even chicks whose parents are present aren't entirely invulnerable: sometimes penguin males accidentally incubate stones instead of eggs, so that after nine weeks they have no chick. No infant penguin can peck its way out of a stone. Thus, if a hatched chick wanders briefly off its parent's feet, disappointed fathers of stones may chase after it, trying to manoeuvre it onto their own feet. It is hard to be tender when you are disappointed, and when you are vying with six other disappointed penguins, and when you have only clawed feet with which to appropriate babies; sometimes the craved reeling penguin chick gets kicked and frozen to death; although one unrealistic adult may still kick its spoiled floppy body along on the ice, still dreaming that it will mount his feet and be his own.

While birds seem as suited to the Antarctic winter as morning glories are to boiling mud, penguins do have some qualities to make a winter on ice sufferable. For example, stiff legs, which make the standing easier. If penguins can stand through four-day blizzards without getting dashed down and exhausted by snow, if they can live through their first year — as two out of five do not — and grow streamy slick feathers and stop looking like fraying bags; and if their parents, finally, when the chicks are strong, stop arriving with squid, so that the chicks become hungry enough to wobble off, through miles of golden blowing snow, to find the unseen sea, then they will discover, once they leap in, that they have talents besides standing – swimming with sudden winging, wheeling grace in water.

OSTRICH NESTS ARE hardly more lavish than penguin nests: a scrape in the sand for their twelve peeping eggs. (The eggs peep so that they can orchestrate hatching day.) Twelve peeping eggs sit on the scrape, where they can be eaten by hyenas and baboons, where lions can roll them away to play with, where Egyptian vultures can drop rocks on them to crack them and stop their peeping.

Sometimes ostriches start twirling, or running in circles on the sand. To what purpose do they twirl? Who can twig the intricated soul of the pirouetting bird? A little is known about the rotary impulse in other creatures, like lost people. Lost people travel in loops; only, they don't know it, because a big enough circle seems like a line; and the flying moon and the shifting moths, and flowers which tip east and then west along the way, further prevent the insight that something is repeating. But are ostriches lost too when they spin? Is twirling a kind of programme by which they are not allowed to travel forward when lost – forward where they might become loster, forward where they might end up among bad hill beasts? Or are ostriches, like many who twirl, trying to forget yesterday?

In any case, her twirling hazards the ostrich her offspring, for if she gets off her nest to twirl, the lions may bound in and roll away the eggs; and if, later on, she twirls instead of attending to her chicks, then another ostrich family may abduct them. Nontwirling ostriches can accumulate hundreds of chicks.

So not many of an ostrich's chicks hatch, and not many of those who hatch live long, and not many of the chicks who survive remain in their mother's possession. As a parent, the ostrich is dippy. And of course, although she flaps her wings joyfully, the ostrich will never fly away, for she is very heavy and her feathers are ruffly rather than aerodynamic. Neither are ostriches good cart-pullers, because when they get tired, they just plop down.

But when she starts to run, then the ostrich laughs at lions, and she laughs at horses and carts and hyenas as she passes them, for she has long pink rifle legs with which to course across the sand.

FROGS ARE TOO PRETTY for anyone to seriously believe that they disdain attention, but many frogs when they sleep try to screen themselves behind leaves and rocks, to hide from people who want to poison each other with frog poison; or people who want to cremate them into frog ash, which, when worn around the neck, foils the plague; or frog-sucking people who wish to baffle themselves, or to anaesthetise an aching tooth, or to forget yesterday by toxifying their heart muscles.

And so, during the daytime in Borneo — when many frogs sleep — the frog in the tree is like a sea bean, floating out where no one can see it. Frogs sleeping all over the earth — mist frogs, tinker frogs, glass frogs, leaf frogs, wood frogs, and frogs from Borneo — are like so many varieties of sea beans, sleeping unseen in the leaves. But when one sort of frog from Borneo wakes up, instead of progressing down the tree, as other frogs would, and shoeing through the underbrush along with sun bears and bearded pigs and barking deer, and then forcing its way up a tropical heath tree strangled with vines — it never minds the underbrush, never minds the sun bear: it extends its legs and flares its ball-tipped fingers, with webs in between, and glides off, from heath tree to heath tree, steering through the air with immense parasol hands. Nighttime in Borneo is lit by pale green glowing mushrooms, which glimmer on the nocturnal talents of the green flying frogs.

THE TAKAHE BIRD in New Zealand has a massive beak for breaking eggs and seizing fish and tearing carcasses. But while New Zealand swamps do have fish and eggs and carcasses for seizing, they also have stoats who eat takahe eggs. When the stoats arrived from Europe, the takahes left their swamps for the mountains and learned about grass. Now they live by themselves in the mountains above the fjords, eating long thready tussock grass with their strong beaks, like a nibbling basilisk with no one to look at. The takahe has turned nibbling and dogged, for eating grass means you have to eat all day.

But the crusher beak affixed to takahes is not only irrelevant. It is also a problem: it is bright scarlet as well as fierce, twice too glorious for life in the grass. The takahe must be aware that her beak is so red, for when a hawk flies over, she puts her face to the ground, or she hides her beak under her wing. A long time ago, the takahe stopped flying, but fortunately she still has wings: if you decide to stop flying, it is wise to hang on to your wings, for you may need them one day when one of your other qualities turns dangerous. So while the takahe may not have the chance to feel anything analogous to what penguins feel when swimming, and ostriches when running, and frogs when flying, at least she has a small wing under which she can hide her talent, lest the hawks fall upon her.

Warbler Delight

The water is wide;
I can cross o'er.

I suppose that someday, suddenly, I will be transferred to another age, for example the chivalric or the bronze. The hope is, of course, that I arrive in period dress but not resemble a contemporary luminary, for I wish to simply onlook. But, more probably, thanks to chronologically garbled garb, or my mistakable face — which will lead to expectations of competence — I will have to explain my occurrence. That explained, I will have to explain my age, The Present, also known as 'The Future' in the past. This is why I am studying our great inventions and advances: to be ready for questions.

First of all, it seems imperative to understand modern bird migration, for birds used to fly to the moon in September and then back in spring. Now, why birds wintered on the moon is a good question, but this is what people realised was happening when they saw swallows flying toward the silv'ry globe. Birds nowadays usually just go to Brazil or Morocco for the winter. Thus I hope to be useful to the exhausted birds of the past by explaining how their posterity succeeds with much-abbreviated trips.

One little bird, however, performs a migratory feat reminiscent of birds' wintering-on-the-moon days: starting out from Alaska, the blackpoll warbler flies three thousand miles east to Nova Scotia. There he gorges himself on webworms and sawflies and gets fat while waiting for a strong northwest wind to blow him off his twig out over the Atlantic Ocean. Thus begins his two-thousand-mile transoceanic flight to Venezuela.

But *fat* is a gross word for a trifle-sized bird — a four-inch-long sprite knit of feathers, hollow bones, and heart. Warblers are not beefy like geese; a goose on your head gets irksome, compressing your neck; but a warbler could spend the week there undetected, like a cherry or a shilling. Even with their enormous hearts, warblers weigh one-third of an *ounce*, which means forty-eight warblers to the *pound*!

They spend their first three months eating insects in the spruce-fir forests of northern Alaska and Newfoundland, staying within an acre of where they hatched in June. But after this rather provincial upbringing, the tiny spirits are gripped with a restlessness to pitch themselves into a sixty-five-hundred-mile trip over unknown terrain, to arrive in an unknown land. Comfort does not fascinate warblers: even if you put them in a warm, wormy cage in Ohio, come September they're still facing impatiently toward Brazil, hopping and scratching and frantic. 'Only Brazil worms!'

Terns and shearwaters also fly astonishing distances over water, but as they are flying *and* swimming birds, the whole ocean is for them a stopover. They can plop down on the water when they get tired and have some Fish Delight. Blackpoll warblers cannot swim, for they have tiny grippy bone-toes that do not serve in the water (try swimming across the pool with a fork in each hand). If they touch down they become Warbler Delight. They are not waterproof and they do not float; they get soggy, then sink. And so they must keep flying from the coast of Nova Scotia to the coast of Venezuela — flying for eighty or ninety hours straight! They do not glide tranquilly either, like albatrosses; they fly like this:

Flap flap flap Drop Flap flap flap Drop Flap flap flap

If they are not flapping, they are dropping. With short wings, perfect for chasing those wickedly nimble Alaskan flies through thick mazy jumbles of spruce and thickety brush, warblers are vigorous and sparkling little flyers, but they do not soar. Daedalus did not build Icarus warbler wings. On long flights they sprint forward, then fold their wings back and drop — many times a minute.

They find their way by the stars, they find their way by the sun, they find their way by little crystals in their heads that orientate them magnetically. They navigate by landmarks, too, landmarks they remember every year — which is why young birds who have nothing to remember might fly to Minnetonka. Migration doesn't always go perfectly, especially the first time around. Sometimes inexperienced ounces flying over the ocean get blown off course and end up in Ireland. It must seem strange then, the whole plan: leaving the cozy thicket, flying halfway around the world to drizzly Glashabeg.

On the beach in Venezuela, the warblers just lie there, feathers and bones, all the worm-fat spent. After they land they lie sprawling in the sand, dazed, following their four-day transcendence. People can walk right up to them and they don't care, they've just crossed the ocean. But then they shake off the sand and the lassitude, they fluff up and eat some spiders and carry on for fifteen hundred more miles to Brazil for the winter – also known as 'the summer' in Brazil. Except for a few Irish vagrants, modern warblers do not winter on Earth after all. We winter, we summer, we winter, we summer; while the warbler flies from summer to summer to summer to summer!

Pea Madness

If you have only one mind, or one cooking pot, you will be forced to entertain a hodgepodge. Many of the pre-Vikings only had one pot, so they ate their peas as part of a hodgepodge called Ärtsoppa, with millet and sesame and panic and poppy. Ärtsoppa upgraded the pea — *any* fellow ingredients would have upgraded that ancient pea, musty and cartilaginous, good for filling silos or baby rattles.

By the sixteen hundreds, however, peas had become dear, like delftware. Many Holland seedsmen, now involved in obscurity, were once involved in the quality of the pea. They rendered the rattle-pea succulent and creamy. Peas were not just averters of hunger anymore: pea-eating was a madness. Fresh tender green peas were escaped from dances for, stolen at midnight from kitchen larders, smuggled by pocket into church, considered one of the very merits of life, often so hardtack.

At first the life of the pea does not seem to be that of a mad-maker. The young pea plant lives by diligent routine, forming two tiny equal leaves every four and a half days. If leaf-leaf on a Tuesday morning, then leaf-leaf on Saturday evening, and leaf-leaf on Thursday morning. Someone who helps peas — a friar or a bee — may look in on them, but young peas are as autonomous as mushrooms and responsible as clocks.

But then, what had seemed a mushroomlike spirit of autonomy turns out to be just the delusive stability of shortness. Peas are clocky children who become spoony adults. Once they grow long-limbed, they start to teeter, because they possess more self than they can support. Then they grow madly wending tendrils, to sweep the air for lattices — just as teetery marionettes will grow marionette cords to sweep the air for marionetteers. Yearning begets yearning: the pea plant yearns for a lattice, so it grows tendrils — then every tendril too yearns for a lattice. Yearning draws tendrils out of the spindly green pea-shoot only to find itself compounded, elephantine.

Tendril wending is swervy and conjectural; like a dancer who cannot quite hear the music, pea tendrils are antic with inapprehension. Since there is no way for them to apprehend a lattice, the only direction

32

to grow is yonder. Haywire personalities like peas, wobbly personalities with loose ends, iffy ends, result not from having no aim, no object in life, but from having an *extrasensory* object. What they want is beyond their powers of apprehension — until they hold it in their acute green wisps — so their manner is vagabond. The personality that longs only for perceptible things is down-to-earth, like a dung eater. But the teetery-pea kind send out aerial filaments to hound the yonder, tending every which way, guessing themselves into arabesques, for they are fixed on the imperceptible.

The truth is that lattices are not the only things that are extrasensory. When you cast your small, questioning arms into the opaque universe, you may find a trellis to tether yourself to; or you may find a tree sticky with birdlime; or a snuffling piglet; or a trapeze artist swinging by who takes you for an aerialist and collects you — then alas, unless you have excellent timing and a leotard, you will be a lost cause.

Or you might find nothing, in which case your yearning will unhorse you. Yearning can horse you or unhorse you. You can only look for so long before your looking apparatus topples you over. Maybe there is no lattice within your reach; for not every plant is issued a lattice, just as not every planet is issued a people and not every person a pudding and not every pudding a plum. Or there may be a lattice right there

33

next to you, installed with you in mind, around which your feelers verge and twist but never touch until finally, freighted with longing, you fold, same as the plant with nothing nearby. The lattice an inch away might be a moon, a myth, an abstraction. You might have grown your tendrils as filigree.

If the road from pea to pea is shortened, the plants can be lattices for each other, like marionettes winding their cords around one another — *hold me up, Blue Baroque Lady; hold on to me, Lindenwood Gnome*. But then the pea patch becomes a pea ward, because 'mildew cometh by closeness of air'. Crowds do not divide the blight but multiply it. On May Day everybody dances around the maypole together and everybody feasts together and the plague travels from him to her to him to him. More people means more plague, more wood means more fire, more peas means more mildew. Pea powdery mildew will infect the pea and the pea and the pea, not just the pea.

Pea powdery mildew spores travel by being puffed around by the wind — so they might get blown into a community swimming pool, or a henhouse, or onto a freshly painted mural. In fact, anything but a pea plant is a cemetery for them; though sometimes they land nowhere and remain in the air, unberthed with the specters. Once in a while a flurry of spores gets blown straight into the thin, seeking arms of a *Pisum sativum* plant. Then the spores settle down and with their hollow haustoria grieve the plant; they drink its terminable green blood, making it powdery-pale and curly-weak; but they drink carefully, sipping slowly through their straws, for how can you drink from an empty glass? If in all this wide herby world, if in all this gardeny world of moonflowers and lespedezas and daphnes and daisies and frangipanis and ghost weed and bluets and galaxes and blazing stars and blood-on-the-snow and mind-your-own-business and porcelain berries and rain lilies and chinquapins and withywinds and salsifies and fritillaria, you light on the one thing you can live on — pea plant — why would you drain it dry and give yourself back over to the air? The air is a question and those who travel upon it travel in questions: When will I find what? Where is who?

LONG AGO, LONG BEFORE the air and its travelers, there was an immense void between Muspelheim and Niflheim, the lands of fire and ice. It was called the Ginnungagap — the gaping gap, dark and empty. But when the Fire Giants from Muspelheim trooped across the Ginnungagap to war with the Frost Giants from Niflheim, there was provoked some lively melting, and the yeasty droplets flung into the void became persons and plants and beasts. These proliferated and filled up the emptiness; and for the ones who can easily reach the materials they need, the Earth is nothing like a gap; they are content, like clocks endlessly fingering integers.

But there are others who still experience the Ginnungagap, like buttercups. Into the gaping gap buttercups send their yellow-dusted anthers, petitioning for a bee; and the gap is where the crane flower sets up its blue landing-petals and orange flicker-flame petals – in order to doubly tempt the sunbirds – with blue stability *and* orange witchery. And into the gap the *Pisum sativum* dispatches its loopy tendrils. But like a lost wolf howling for her pack, whose long strains of *find me* might attract the wrong wolves, the plants might summon the wrong thing with their susceptible anthers and petals and tendrils. So instead of sunbirds the landing-petals could receive Diseases that cause corkiness, mushiness, yellowness, sogginess, stinkiness and spots.

Why even take the chance? Why try at all? Only soaked in full-cream milk and putrid egg yolks and train oil and copper arsenic can anyone be invulnerable. So why not stay safe in the dirt, a seed holding tight, instead of a seed bursting forth and offering the plummy stationary self to slime mould and powdery scab and blossom end rot and weevils and sow bugs and gangrene and silver scurf and scrappy little sparrows waiting above ground?

Plants cannot stay safe. Desire for light spools grass out of the ground; desire for a visitor spools red ruffles out of twigs. Desire makes plants very brave, so they can find what they desire; and very tender, so they can feel what they find. Thus genips with hearts of honey-pulp; thus poppies with hearts of fringe, and pickerelweeds with hearts of soft pale purple frill, and tulips with tilting hearts, and foxgloves with downy freckled hearts, and the maddening-sweet hearts of the careening pea. Those who are feeling their way into the Ginnungagap must be able to feel, which means able to freckle, and fringe, and soften, and tilt. And if they can tilt they can fall; which is a different design from that of the ticking hearts of crystal-quartz.

Radical Bears in the
Forest Delicious

There once was a king of Babylon who was too proud, so he was given the mind of an animal and put out to pasture. For seven years he roamed the fields on all fours and munched on grass, after which period he was allowed to return to his palace and rich robes of purple, his barley beer and skewered locusts and royal hairdresser who gave him back his dignified ringlets. (Along with an animal's mind he had been given the animals' hairstylist.) It is not specified which animal's mind Nebuchadnezzar received, but from his glad return to civilisation and fine cuisine we can infer that it was not the mind of a panda bear. If he had had a panda's mind for seven years, in the end he would have rejected the restitution of his kingdom; he would have somersaulted away, to continue leading a free, elusive, unfollowed life.

Having followers is an honour pandas dream not of. There is no tragopan so trustworthy, no bushpig so dependable, that they would want it tagging along. Pandas even head away from pandas, like the stars in the universe, spreading farther and farther apart (you can never be too far away to say goodbye) — except their territory is neither infinite nor expanding, and in order to deliver more panda bears into existence, they can't just scatter into particles at the end. Pandas come together every two years or so; marriage isn't always marriage of the mind.

Maybe if they had been given a choice they would have picked a less conspicuous coat, one to better correspond with their reclusive spirits. Admirers can be secret admirers and afflictions can be secret afflictions but pandas cannot be secret pandas, since they contrast dramatically with green ferns, grey rocks, pink rhododendrons, and their own bellies and ears and legs. They are showy bears, sensationally visible, which might actually be an advantage for a solitary species: the easier to avoid you, my dear. Camouflaged animals must always be bumping into each other.

What does the animal do all day who is not engaged in society, its duties and pleasures and ferments? There may be some wedging in trees, some gazing into the mist, some fiddle-faddle. Sometimes the panda breaks an icicle off a branch and tosses it into the air over and over till it melts. Sometimes, trotting pigeon-toed across a hillside, he trips, then rolls, because he is round; having enjoyed that, he climbs back up and rolls back down. He might pick wild irises or crocuses and recline among the fern-fronds to eat them, or lounge underneath a weeping willow, munching on the little leaflets that dangle into his mouth.

Mostly what pandas do with their time is eat bamboo. Bamboo, that sturdy wooden grass, makes up to 99 per cent of their diet and they eat it for up to fourteen hours a day. They have to consume it constantly since they are only assimilating about 20 per cent. Their penitential diet is a mystery; pandas are like celery saints — everyone else is convivially dining on stuffed eggs, truffled fingerlings, little pies and oranges, enjoying the tableside crooners, while out behind a bush sits a celery saint with his basket of celery, crunch crunch crunch. Eat enough pies and you can put aside the desire for food and pursue something else, such as a cowhand. Rare is the romance of the celery extremist.

With their carnivorous anatomy and herbivorous behaviour, it is as if pandas are pledged to an ancient covenant — as if they used to be bon vivants like other bears, blood and berry juice staining their muzzles, slugabeds all winter, until one day they fell into a trance and received a deep message: 'You are standing, pandas, on the very borders of the eternal world, but you have become charmed with infatuating food; the subtle poison of sensuality courses through your veins. You must disregard custom and the strong clamouring of appetite and passion. It will take, at times, every particle of willpower which you possess; but give yourselves wholly to a bamboo diet, and guided by firm, unspotted principle your lives will become pure and noble.' Thus was formed that radical sect of bears, the Bambooists. Modern-day Bambooists show a remarkable resistance to temptation: a stream runs by, serving up fresh fish, and what does the panda do? Wades across, to get to a stiff thicket of bamboo on the opposite side.

But willpower might not entirely account for such abstemiousness anymore. Bamboo is not power food, and the bear who eats it is not a power bear, and swiping fish from the river takes energy, as does sleeping all winter. If you're going to sleep for seven months you need to eat your hickory nuts, your ungulates, your honey. Bambooists have to stay awake all winter to eat bamboo – incidentally witnessing the sapphirine sparkles of snow falling from a branch, the cliffs draped with icy fringes, the white snow powdering the green bamboo leaves. (Could any dream compare with winter?)

What does a panda know, who studies just a few cloudy-mountain miles of the world? From her experience she must know about fallibility. Icicles melt, flowers fail, intangibly small babies grow tangible and autonomous, and one day when you come back from foraging to collect yours from the tree fork where you left him, he is gone. Mushrooms, moonlight, everything is ephemeral, with one exception: bamboo. Bamboo never fails, bamboo is eternal, evergreen, green in the orange season, green in the white season, green in the green season, poking up sweet little shoots into the spring rain. Blessed is the bear that trusteth in bamboo.

For lucky pandas it is true, bamboo never fails. Bamboo can be eternal for a hundred years, which is four times as eternal as panda bears; but there is in the character of bamboo a devastating defect. Most grasses stagger their dying, piece by piece, like an orchestra — though a trombonist goes down the collective life carries on. The trouble with bamboo is that it crashes all at once: after a century of continuous availability the entire thicket flowers together, dies together, and like a dead orchestra it can take twenty years to get back on its feet.

At this point an animal might wise up and become a Whateverist. With so many edibles in the world why consume, almost exclusively, a miserably nutritious, erratically fallible one? It's not as if bamboo is pleasant to eat, like horsebeans; bamboo splinters poke and scratch the swallower all the way down. That old covenant was arbitrary and perverse; bamboo is a silly staple; specialism is folly. Consider pragmatists — when the linguini runs out, a pragmatist will eat the centrepiece, and when that is done he will eat the tablecloth. As pragmatists have no principles their numbers are myriad.

But pandas betrayed by bamboo go looking for bamboo. For there is such a thing as specialised hunger, being hungry for one thing — similar to specialised loneliness. Sometimes they don't have to travel far; pandas eat several kinds of bamboo, and even though arrow bamboo collapses, there might be umbrella bamboo growing nearby. Sometimes they have to go farther afield, and sometimes they travel in pitiful directions — would you know which way to go to find a hotbed of celery? — until their coats don't fit very well anymore. Vagrancy used to be easier on the animals, because there used to be more forest. Even if an expedition wasn't efficient, it was foresty all the way, just as the journey from earth to heaven is milky all the way. Now, between patches of forest, there are villages and gravel mines, steep cornfields, dance tents, frightened people waving blankets, mushroomers, other things to avoid.

People have tried to help pandas become pragmatists, to see sense, to switch to alternatives during a
bamboo strangulation. And in captivity they comply
— they eat the yams and bananas and fish set before
them. But compliance is not conversion. When they
are set free, pandas return to their ruinous fidelity to
bamboo, shuffling past opportunity — for on the far
side of that hill might be the Forest Delicious, where
they can lie back, in the million-column sanctuary, a
bamboo cane in each forefoot, crunching on the one
and then the other, munching on flappy bundles of
leaves. There are fewer than twenty-five hundred free
pandas left and they're all in the same boat, made of
bamboo. When it goes down they go down with it, into
dark water, and they won't switch to another boat, not
for all the tea in China. Pandas have their own wisdom, unaccountable and unamendable, whose roots
shoot down deeper than we can penetrate, and if they
mind anyone at all it is someone more elusive than
man.

Trooping with Trouble

To whom, then, does the Earth belong?' said the dragon as he was being slain. 'Sometimes it seems to belong to dragons; at other times to dragon-gaggers. Sometimes it seems to belong to the harmattan wind, then to the doldrums. Sometimes to the slaves, when the sea parts to let them through, and sometimes to the sea when the sea does not part. Now to the siskin finch and sablefish; now to smitheries and smelteries. Perhaps the Earth is neutral, like a bridge between two cities, travelled on but possessed by no traveler.' Such are the behindhand ponderings of a doomed dragon.

Turtles do not have to be doomed before they become canny. While it is not known how much eggs understand, the tiny sea turtles nosing out of their leathery eggshells recognise instantly that this is not a turtle's world. A hatchling who believed otherwise would scramble out of his sandy burrow and, upon emerging, stop: 'The world is my sphere and the sun is my travelling lamp and I, soft-shelled, shall span my sphere and nothing can ruin me.' Then he would step sportily down to the sea.

But in fact the hatchlings unbury themselves gravely, stealing up out of the sand over several days, to give their soft shells time to harden. Once they are close to the surface, they pause, and wait for the sun to go down; then they run frantically down to the water, and beat through the shallows to get to the deep. The sandy newborn turtles shimmying down the Yalimapo beach as night falls are no ninnies. The dog, the raccoon, the wave, the catfish, may ruin them and there is no mother.

Neither does the world belong to snakes, as they forthwith apprehend: stuffed into clay pots by snake-catching guilds, vipers are ignominiously milked of their venom, which is then turned to antivenom. Potted snakes employed in cancelling their own powers cannot consider themselves sovereign. Even the king cobra, fifteen feet long, finds itself ridiculous when giddy women kiss it lips-to-lips to amass fertility luck for themselves.

Although lizards have *some* edible subordinates, like mealworms and sweet potatoes and snails, lizards *are* edible subordinates, too, for birds and bobcats. The worst menaces that lizards can perform are just medium. Only two — *Heloderma horridum,* the Mexican beaded lizard, and *Heloderma suspectum,* the Gila monster, made of black and peach beads — carry poison in their bites. But while snakes store their venom in the upper jaw, whence it flows into ankles and guild jars, the *horridum* and *suspectum* have their venom glands in the lower jaw, so that when they bite, the poison must seep up weakly through the bottom fangs. Only the seediest victims get bitten long enough to be defeated by the *Heloderma suspectum*; the plummy ones get away. The peach-bead monster can barely access its most devastating ingredient, and its status on Earth is Vulnerable.

Maybe vulnerable creatures — Gilas, sea turtles, eyelash boas and banana boas, mutable tulips, muddy-footed panda bears, muddy-rooted hazels — are more interesting than invulnerable creatures. What do the invulnerable have to gain from invention or experiment? In form, lizards are as experimental as pasta. There are flying dragons with ribbed side-flaps that fold out like wings. There is the thorny devil so wide it can eat thousands of ants at one meal. There are whiptails, all of whom are female; there is the Jaragua sphaero, a tiny gecko living in leaf-litter, as little as geckos can be without drying out (that tiny,

47

it has a dangerously high proportion of surface, and little interior to keep it moist). There are the lizards with crizzled necklace-frills and the Jesus lizard who runs on water.

Then there are the saggy chuckwallas. Simulating largeness requires skin saggy enough to not tear when expanded. When they are hunted, chuckwallas, polygamous vegetarian iguanas of the Mojave Desert, wriggle into a cleft in a rock and then inflate. The bobcat would like the chuckwalla in its throat but instead the chuckwalla is jammed in the rock's throat.

Other lizards hide by crypsis, or blending in, like Neptune keeping secret among the stars until 1846. Some lizards look like leaves and some like tree trunks and some like thorns and some like beetles. The secret to crypsis is placing yourself among things you look like, but in a scene where no one will expect you, like Willie Nelson with Lithuanian peasants.

It is also possible to be cryptic by just being drab; then you can go anywhere without being seen. The problem is that females of your own species will not see you and you will have no offspring. It is best to be drab or brilliant as needed, like the cyan-tailed skink, which practises *sudden crypsis*. To perform *sudden crypsis*, one needs an easily sheddable pretty part, like a yellow-and-green kite or a tangerine ruffle skirt or a bright blue tail. Fly the kite merrily; wear the skirt, the tail, merrily; but when you sense you are being hunted, dispatch your accessory, and then just as merrily be drab, for your hunter will swoop after your castoff and you will be safe and invisible. Once untailed of its electrifying blue, the skink is brown and blithe and more essential.

Perhaps animals that troop with trouble have more character than animals that are above it. He who needs not respond to the bobcat or the wave or the coyote needs no frills or side-flaps or soft speedy feet or detachable blue tail; he needn't be wedgeable into rocks or mistakable for thorns. He can reach his senescence unaffected by experience, not made gracefully peculiar.

The Jackson's chameleon is an accidental transplant from Mount Kenya to Hawaii. (If, when your chameleons arrive in the mail from Africa, they seem kind of dry, letting them onto your lawn is not the best way to refresh them: they may disregard your fence and populate the mountains beyond.)

Jackson's chameleons are born falling, in their embryonic sacs, from on high, where the mother in the tree is emitting her babies. The sacs break when they hit the ground and the chameleon babies burst out, a little shaky. Then they climb a tree because they have tree-climbing toes. And long tongues for drinking rain dribbling off their noses, since puddles do not form in treetops. Also independently moving eyeballs for watching groundward and skyward at the same time; and flexible complexion; and three long facial horns, good for confronting relations with horns.

But the Jackson's chameleon also confronts *itself* with its horns. Sometimes as it is roving the tree branches, gaping and hissing and swaying and surprising wasps with its projectile-tongue, it will by mistake grab onto its own forehead-horns and then panic, wrestling itself, frantic to escape its own frantic grasp, a one-reptile bedlam in the paddle-leafed tree. As the sunbeams glow across the blue seawater and web through the forest of pale yellow bananas and green bottlebrushes and pink-orange guavas and mix their tints around the thrashing chameleon, it can shift its hues, to match the ambient light, from green to turquoise to sour green to blond to blotchy olive-blond. But such wars make such graces ridiculous. The problem with being prehensile is that you might seize your self. Thyself may be thy chiefest rascal.

Silly Lilies

Most plants will bend over backward to cooperate with reality. If they live on a shelf far from the window, they will bend and warp toward the light, as you do when you are sitting on the living-room sofa and an intriguing stranger keeps wandering through the laundry room. Or if they poke out of the dirt and find themselves buffeted by windborne ice crystals, most plants will modify their dreams of stature. Stature is just gravy, dreams are just gravy, everything but the ghost is just gravy. So long as you hold on to your ghost it is fine to stoop.

But sometimes the odd plant, like a mouse-ear cress, goes batty-bat and bends over backward to co-operate with *unreality*, sending its leafy shoots burrowing into the ground – as if the sun were down there somewhere, you just had to keep trying – and its roots rising up, as if phosphorus blew on the breeze.

Anyone might lose his way, of course; anyone might stumble during the torch dance and stifle his own flame. You could try turning the little plant over — *your roots should go down and your shoots should go up* — for, once advantage has been experienced, will it be rejected? Once they taste its rich peaty drink, won't the roots want to stay in the dirt? But this is not a garden-variety plant but a gravitropic mutant. In response to gravity, a worldly plant sends its roots down and its stems up; this plant does the opposite; it is like a demented boat that insists on sailing upside down, draggling underwater its silky sail. As many times as you replant the plant, the plant replants itself, uprooting its roots, weaving its tiny leaflets back into the dirt, looking for the sun underground, looking for disappointment.

In Crocodilopolis the young blue cornflowers alongside the road are manifesting in the worldlier way — roots down, shoots up — and feeling in fine fig: 'I have my mother's petals!' 'I have my father's filaments!' There is no ice in their future, nor dementia, nor need they warp and buckle to apprehend the sun. The sun is not a mysterious stranger for them like it is for some houseplants. Rather, the sun seems intrigued by *them*, even enamoured, so ardently it sheds its light on them! Oh dazzling blue, oh radiant blue flowers! Perhaps it would like to spend the night with them too, instead of always continuing on to the Strzelecki Desert; but the sun can never spend the night.

A few of the cornflowers will get plucked for confetti, a few for potions to soothe tired eyes, and a few to reanimate the dead. All of them will animate some honeybee or another. It seems the perfect combination of realistic disposition and congenial reality.

But what these easy-hearted, confetti-petalled flowers do not foresee is that the sacred crocodile will succumb to elderliness this year and there will be a terrible, trampling procession to honour him and a terrible, trampling procession to instate a new sacred crocodile. In their enthusiasm the Crocodilopolites will tread on the flowers that were not pursuing disappointment. (Unlike skunk tramplers, flower tramplers are oblivious.) Few flowers are trample-tolerant; in crocodile succession years it is better to be a flower in the pond than a flower in the way. Even the most enthusiastic worshiper will remember to circumvent the pond.

Far from the sacred commotion, water lily rhizomes lie in the mud at the bottom of the pond, gathering their powers. Complexly for underwater rhizomes, they are buoyant: try taking one of the buoys that delineate your swimming rectangle in the lake and swimming it down and burying it in the sandy bottom. Lily rhizomes develop retractor roots to pull themselves down into the thick mud, to squirrel themselves away; but such buoyant natures are at risk for untimely launching, like an astronaut who leaves without his rocket. If they are not buried deeply enough, the knobbly bulges pop out of the mud like volunteers and float on the surface — still lilies, but lilies estranged from their vocation. Many latent florists express themselves in other ways — basketry, for example, or banditry — but lily roots rinsing around on top of the pond cannot even serve as rectangle perimetry, for soon they get soggy and disintegrate and sink. Sometimes repression is good. If I had stayed in my egg I could have been a dangerous green dragon.

Lilies *are* supposed to ascend from the mud, but less like underprepared astronauts and more like someone with a backpack full of rungs and planks who assembles her ladder as she climbs, rung by rung. Riding on top of this climbing assembly is a large leaf, double-scrolled like a scripture. As soon as it reaches the surface it unscrolls, gladdening green darners and hopper nymphs. The Day of the Lily Pad rejoices many tired and aimless organisms: snails, sponges, water scorpions, springtails and pill-clams, damselflies and biddies and backswimmers, islanders who haven't had an island.

Starting on the Day of the Lily Pad, water lilies draw fresh air through tiny stomata on their leaves down into their hollow rhizomes. They breathe unconspicuously (unlike accordions). But a plant that can breathe is a plant that can drown — lily pads are not splash-tolerant. If a fountain should immigrate to their pond, and they never have a chance to dry out, lily pads turn brown and droop and dissolve. Then the islanders have to go back to circulating.

Lotuses drown like ducks when they get splashed. Their leaves are water-repellent, so water rolls off in little pellets. Hurl thine anvils into the pond, thy cannonballs, thy giant boulders: the lotuses will never get wet! Also, lotuses keep building their stems high into the air, instead of relinquishing their rungs the minute they reach the surface — which is what lilies do (silly lilies). Lotus leaves hover up to six feet above the pond, like magnificoes. That is why lotuses are highly flood-tolerant, and why more hippopotamuses doze in lotus lagoons than in celery bogs. Under the ruffly round green canopies the hippopotamuses can lie low, feigning absence, or they can trundle around like gigantic inklings. They can protect their pink tender ears from the sun that fireball. The only trouble is that in a windstorm the gently swaying parasols turn thwacky, like when the hula girls around you start slam-dancing.

After the squall the hippos are bestrewn, like passionately celebrated sopranos, with petals stuck to their foreheads and greenery embellishing their rumps. They resemble the little remembrances Egyptians used to take with them when they went to eternity, those little blue-green pottery hippopotamuses decorated with lotuses and lily pads and grasshoppers and rushes and reeds.

But hippos recover quickly from all the whap-whapping and the decoration, and what with their goodly build they are wind-tolerant, like mudstone and mastodons and Zemuzil, duke of Pomerania. Happy is the heavyweight, he shall not be overthrown. As for lotuses, they are *not* wind-tolerant; their stems are hollow and slender like magic wands. Though you can use your magic wand however you like – you can access other realms, or deepen someone's slumber, or pick a perfect pumpkin (ripe pumpkins make a hollow sound when thumped), or wave it back and forth until a heart-stopping pink-orange flower appears on the tip, or rap it on a music stand to beat time for the neighbours who like to sing together in the evening, or hector a hippopotamus with it – magic wands are breakable and last longer when they are used for magic.

Megafauna are not the only ones to survive a windstorm, though. Water lettuce also does well, because it does not take a stand; it has loose and dangly portable roots. The same windy thrashings that deflower the lotus and flower-flog the hippopotamus and send rock-a-bye babies tumbling down from the treetops, feel to the water lettuce like the teaberry shuffle. It swivels, it bounces, it shakes its flouncy leaves, it skims around the lagoon like a passing fancy. And because it makes new daughters wherever it goes, water lettuce can never be in exile; its terminus is immaterial.

But the season is not. Water lettuce is wind-tolerant but not winter-tolerant; unlike *Veronica beccabunga* hunching underwater or swamp candles standing up to the cold, water lettuce freezes hopelessly. Winter is its bitter end. Imaginably, a water lettuce could catch a stream south to Florida and overwinter in Sarasota like a circus; but water lettuce is illegal in Florida, so it would have to somehow elude the authorities. Oh isn't it always like that! – wherever you are immortal you are also illegal. In another overwintering scenario, a cottager might take the water lettuce in and float it in a bowl in the south window and hang a lightbulb six inches above it and remember to treat it regularly with muriate of potash. But most people feel pessimistic about their ability to undertake such a project.

If they are going to have a dependent in a bowl, most people prefer to grow a miniature lotus, since its flowers look like little pink Yablochkov torches. As with all lotus flowers, after the petals fall off, a platter-like seedhead is left, fitted with a socket for each seed. The platter holds itself level for several days, then tips over and spills the seeds out. Some seeds will drop back into the rice bowl and go on to live happy containerised lives. Some will bounce out the window and go a-blooming in the backyard puddles. Others may get a-pounded with sugar and peanut oil into a sweet paste for stuffing into porcupine-shaped buns. A seed or two might accidentally fall into Lake Baikal, the deepest lake on earth. Rice-bowl lotuses sprouting at the bottom of Lake Baikal will try very hard to cooperate with reality; which will mean *not* stooping or warping; which will mean stretching up as tall and straight as they possibly can. They can grow up to thirteen inches tall, but that will not be tall enough to reach the surface. They will not be able to open up their parasols or their torches. Try climbing to the Moon with only thirteen rungs in your backpack.

One seed might fall into a dried-up well and lie there for a thousand years. Lotus seeds have such hard shells that they can wait that long to germinate, until the time is just right. What does it feel like to be a dormant dwarf jade oriole lotus with delicate pink petals? Maybe it feels like being a dormant swamp candle, or a dormant giddy water lettuce, or a dormant gravitropic mutant doomed to thwart itself in a salubrious world. What does it feel like to be dormant at the bottom of a dusty well? Maybe it feels like being dormant on Neptune, where the winds blow at twelve hundred miles per hour, or on Mercury, where the sun is less like a friend than an arsonist.

What does it feel like to be dormant for a thousand years? Perhaps it feels like being dormant for a million years. Or for the twinkling of an eye, while a thousand generations of bright blue cornflowers flower and get plucked for twisting together with pale blue water lilies, for draping loosely around someone's shoulders, for weaving into his hair and slipping between his thin fingers, to serve — along with a blue-green hippopotamus, its back legs broken so it doesn't get too boisterous on the journey — as remembrances of Earth.

The Wine of Astonishment

Not all of the insects you might see in the water are water insects. Some may be bamboo insects that fell off their bamboo or shore insects that got washed off their shore. Environment is not conclusively defining because it may be incidental; the person at the organ is not definitely an organist. The secret to identifying any creature is to note its proficiency as well as its terrain: the same insect, clever on a biscuit, may be stupid in a puddle.

Whirligig beetles can circle, scoot, glide, and swivel on the surface of the water. They can flounce and swing around and travel curly trajectories by beating their four hind spoonlike feet up to sixty times a second. They do not technically whirl; they row in little loops; for to whirl requires either the ability to twist or somebody to spin you. Ecstatics are spun by the spirit and seals with their flexible spines can spin themselves; but beetles can only loop, like watermelon seeds with very small rowing teams.

Because the surface of water is slippery as well as strong, it is for the whirligigs an ideal platform. They would not be whirly on the surface of dough or whiskey. They would be sticky, or sinky. But on water the whirligig is extremely proficient: besides rowing with its four hind feet it can scull with its wings, pivot round one long forelimb or one hard wing-protector, slide trackless over the water or twitch hoops of radiating waves onto the surface to locate crustaceans and other meals.

Whirligigs are secondary consumers, meaning they consume other consumers like mosquito larvae and daphnia. They seek their prey by echolocation, which is different from echolalia, a condition where you helplessly repeat everything that is said to you. Bats locate flies by echolocation, sending out sound waves by clicking; when a wave hits a fly and comes back the bat can tell where the fly is and if it's succulent or brittle. People can echolocate too: they can echolocate vast canyon walls. Canyon walls have echolalia.

Whirligigs do the same thing as clickibats, only with water waves they dispatch by percussing the surface with their abdomens. (However minute you are, your waves will expand.) When the waves return, having bumped into daphnia, the beetles sense them with their fine slight antennae and go whizzing after the tiny crustaceans. Without their feelers they would be as oblivious as bath toys: once, an experimenter removed the antennae from some whirligigs, whereupon some swam vigorously into driftwood, some swam vigorously onto shore, and some swam vigorously into swans.

By gathering in consistent places in the pond whirligigs foster an understanding with the local fish: stay away. When they are disturbed, whirligigs exude a smell like apples, which disgusts fish. So if the whirligigs are always in one place, the fish learn to avoid those quarters altogether. But early in the morning, you will not find the whirligigs in their designated areas; instead, all over the pond, strings of them are sliding quietly along, single-file processions getting longer and longer as more of them join, like stars filing back to the constellation just before dusk.

Where do the whirligigs go at night, that they have to reconvene in the morning? Though they dwell in two-hundred-member brotherhoods during the day, at night they scatter off to be alone, for the night obscures them from predators. Perhaps in a fishless world two whirligig brothers would ever increase their interindividual distance until one lived on Lake Here and the other on Lake There.

For almost all their lives whirligig beetles are found in the water – as cylindrical eggs stuck on the underside of waterlily leaves; as fringe-gilled, worm-shaped larvae stalking underwater food; and then as rowboat adults. As pupae, though, they hang in the air, fastened to shoreweeds by abdominal hooks. They were fastened there by themselves in a former stage and are secretly getting dressed in hard heads, adamant mandibles and accessory jaws, putting on thick wing protectors to protect their true wings.

There are two other times when whirligigs are not in the water, though. If their pond dries up they will grudgingly fly around until they find a wet pond. And when winter looms and water turns steely – unbending, unconducting of micrometric waves, terrible terrain for rowboats in general – they dive down to overwinter in the mud. They become like the other indifferent creatures down there, all of them swept off their disparate branches in the intricate tree of taxonomy and scrambled together in the sludge. Special feelers void, true wings vain, paddly feet stiff, whirligigs seem entirely proficient at inhabiting this domain.

But besides proficiency and terrain, there is one more thing to consider in identifying a creature, and that is the season. Some of the animals in the mud belong in the mud; some are just holding their horses. Best to wait until the vernal equinox to decide who someone is — after sun crosses equator and rain softens ground and warmth splinters ice and impassiveness, and up from the mudbed swim the shiny black beetles, up through the murky pond and onto its gleaming surface. Turning ice into water is like turning water into wine; after that many months of mud, whirling on the starry-eyed water must feel like drinking the wine of astonishment.

When Trees Dream of
Being Trees

The tree decided to stop growing after it grew its thousandth leaf. 'No more,' it whispered, and started throwing flimsily attached twigs and old nests down, and shaking the birds out. 'I am a terrible tree! A thousand leaves is more than enough to prove that! I am slow and slight and my leaves are not lustrous. I have never made a flower, never made an apricot, never made an acorn. Go away birds! I am an impostor tree! I will be a post, if I can just shake off these redundant branches,' and the tree bounced up and down, twirled violently, and tried some catapulting manoeuvres in an effort to fling off its limbs. Nothing much was flung, except for some leaves and a butterfly, and they were instantly free from its flinging force, and ended up drifting away instead of zinging through the air. And so the tree started to slam

itself against the earth. Its branches were most certainly broken this way, but they were not broken off:
such fibrous material does not easily come loose, does
not easily separate from itself. So the tree was hung
with broken creaking branches. Aghast, it felt itself
growing. And, knowing it would only grow more of
itself, it cried, 'I must get out of the sunlight! I must
get out of the rain!' It tried to sink into the dirt. But
trees with their spreading root systems are even harder
to push down into the dirt than they are to pull up.
So the tree finally just stood there with its smashed
branches, exhausted, in the late afternoon sunlight.
The other trees around regarded the tree going mad
without much comment. They had seen this dread-
ful thing happen before, when trees dream of being
trees.

Love

Love-in-Idleness

In the year 3,000,002,012 the Andromeda Galaxy may collide with our Milky Way. At first this sounds miserable, like a collision of two bird flocks. But galaxy members fly farly, not tip to tip. In a galactic collision the stars do not actually collide — as with crisscrossing marching bands, only the interstices collide. (Oh to be like a galaxy, to mingle without wrecking. But then we would have to be composed of so much more sky.) The spaces between stars are so wide that thousands of galaxies have to converge before the stars will crash.

But, unlike the gaps between clarinetists, a galaxy's gaps are sometimes flammable. Our encounter with Andromeda will be a predicament after all if the meshing gaps explode. It will feel like being threshed by fire, and it will disturb the Earth's biosphere – the rummaging turtles and yawning bunnies and cherry-composing trees. Afterwards the Earth will plod round its normal route but gaunt, irrelevant, having lost its beautiful flocculence, its beautiful freight.

But what with all those rich, though knocked-apart, ingredients, some rematerialising should eventually happen. And which of the Earth's biospherical items might return first? Not the ones that were tinkered into existence, for there will be no tinkerers in the beginning. Ingeniously small, ingeniously fluffy dogs, for example, can be expected to take many centuries to come back, as can the hanging gardens of Babylon. Derivatives not only take a long time to derive, but also, once derived, they usually require husbandry. Even if a little fluffy dog did appear without being invented, there wouldn't be anyone to shampoo it and rub its tiny teeth with gauze and trim the silky hairs from between its toe pads. It is hard to walk on silky feet, the little dog would slide, and who knows into what – a tar pit? Similarly, without any husbands the hanging gardens would soon be a hanging matted mess.

In the contest between the cultivated pansy and the pansy's scruffy foreflower, the wild viola, for which will soonest reappear on the blasted planet, the viola, also known as love-in-idleness, will surely win. Even with their faces so fat and cheeks so bright, pansies have the more nervous constitution, the constitution more like bubbles. They are subject to pansy sickness and unless regularly tufted do rot and snap at the heel. Pansies, therefore, will need the world to be prepared with responsible tufters before they come back. But violas — though dismissed in our windowbox age for being too indiscriminate-growing, too easy, for having lax standards — the bumpkin habits of the violas will give them an advantage on a world that's lost its windowboxes. Love-in-idleness may even anticipate the thrips and the bees and the pluvia moths, for although these exploited hexapods are prerequisites for windowbox-flowers, love-in-idleness is a self-seeder.

So when the belated magpies are reassembled, and the belated stoats, and voles, and cotton-bolls, there will be no reason for them to be melancholy and back-looking — they shall be high-hearted from the start, for green will already be at large. Love-in-idleness, that frivolous, haphazard, twiddly weed, that untutored flower, that plant heretofore regarded as Not Serious, will already have threaded the Earth with revelry; for as it grew uncritically before the collision, so will it grow afterward.

What would spring forth on an irrelevant Earth but a plant that had been happy on irrelevant earth anyway — hedge bottoms and snowy boulder fields and sloe-shrub patches? What might venture onto a mothless planet but a flower that had preferred steeps and flats, norths and souths, dry badlands and soggy sumps, springs and summers and winters and falls, a flower that had taken root wherever it found a sliver of dirt, even slurry-soil in a slit in soft green rock, as if it just wanted to grow, grow, no matter where? Other flowers are pinpoint flowers; love-in-idleness is miscellaneous. It is not orderly like ice but mazy, veering, like melted water.

And so, when the Earth begins to undishevel after the Andromeda undoing, love-in-idleness will be the first to return; along with moonlight. For love is as idle as moonlight, and has as many prerequisites.

Love-in-a-Mist

Love-in-a-mist frequently gets planted next to spectacle flowers. Love-in-a-mist has spectacles too — flowers blue like jungle butterflies — but it blurs them under fronds. Extracted from their mist of frond-wisps, *Nigella damascena* flowers resemble triple-twining paper stars. However, a *blurred* blue flower looks like any other blue blur: love-in-a-mist's flowers could be blurred racquetballs, or blurred poison-dart frogs, or little unventilated heretics huddling in the foliage.

Being reticent, then, generally green-looking, love-in-a-mist is a perfect Nearby Plant for ostentationists. If you are an orange slipper flower or a fuchsia knapweed or a ruffly silk-white peony, then you will want love-in-a-mist as your Nearby Plant. You can be a jewel and it can be your setting; and in return for its assisting your allure, you can exhort it every morning: 'The greatest need in the world today is the need for modest flowers, meek flowers, flowers who will abnegate regard and let green leaves overlie them.'

Because it shrouds its prettiness, love-in-a-mist allows people to regroup while they are between-peony. When we look at ruffly peonies all we can think about are ruffly peonies; the same with pininanas. If not for an interpolated margin plant like love-in-a-mist, our thoughts would be inane like sneezes: 'Peony! Peony! Pininana! Peony!' Peonies ruffle their petals excitingly; love grows in a mist and lets us meditate on peonies.

Why did love become love-in-a-mist, why did it efface its posture? Was it truly in deference to the flaunty flowers? Is love-in-a-mist not rather a commemoration of Friedrich Barbarossa, drowned in the Saleph River, his drowned-blue face surrounded by the weedy-green tresses of the siren who dragged him under? Or is the blooming behaviour of love-in-a-mist simply a result of its experience, same as the experience of many loves grown diffident, blooming but hunching under screens? Not all loves and not all flowers are granite-petalled. Love-in-a-mist's flowers may have decided to avoid the open as swimmers decide to avoid the snake pond.

Once, we shipped love-in-a-mist to space to observe the effects of cosmic conditions upon it. *Terrestrial* conditions, it seems, had caused love-in-a-mist's blue flowers to retreat beneath a feathery green cloud of foliage, for on Earth, exposed flowers get sungrieved, rain-frayed, frost-brittled. But space is kind; space greets papery blooms of love not with petaltwisting kicks of wind but pacifically.

When it returned to the Earth from the universe, our love-in-a-mist plant was unaltered. Maybe if it had stayed up there for more than a week, love-in-a-mist would have mutated into love-out-of-the-mist, its stellated sapphires rising from their green diffidence and regaining Aplomb and Posture. Then again, given centuries in the tranquillity of the cosmos, love-in-a-mist in space might only ever become love-in-a-mist-in-space, having suffered too many smarts on Earth to ever lose its huddle, to ever dispel its mist.

Love-Lies-Bleeding

Flowers are commonly known for having pink sprees or yellow sprees and then sticking the following week out, before they revert to stems, sere, unvisited except by munching deer. Ackee flowers, for instance, spend all their glow at once and thence proceed to fruit duty and seedhead service. But the tassel flower is ruby-red as a bud, rubier-red as an adult, and rubier-rubier-red from then on.

When planning a tasteful garden, as when planning a tasteful robe, one places the tassels on the edges. Thus the crimson tassel flower can be seen growing on the borders of bundleflower gardens and gardens filled with toadflax and grindelia and green-winged meadow orchids and agastache and holy ghost and Iceland moss.

The tassel flower, also known as love-lies-bleeding, is assigned to garden borders not only because it is tasselly but also because it can be trusted to stay where it is planted, to not redeploy itself. Love-lies-bleeding accepts the purposes of Horticulture. In contrast, if you decide to border your garden with fairies or bluebirds, they will escape their appointment as only the rootless can. There are in truth two ways to escape: the bluebird way and the dandelion way. Bluebirds defect, like bubbles and luck. Dandelions, on the other hand, escape like secrets: sprawlingly. Plant a dandelion border around your garden and soon your garden will be a dandelion fair.

Love-lies-bleeding escapes in neither fashion. It stays in the parish where it's put, using its energy not to strew about and hold a tassel exposition in your garden; nor to decamp; but to bleed. Now most bleeders who devote their energy to bleeding end up paler than when they started. Fringillines and phalaropes, if pierced, fleetly expend their several teaspoons and then are bankrupt like broomstraw. Small fringilline birds are so soon sapped that within minutes of being punctured they can only dream of the days when they had blood: fringillines-lie-dreaming.

There are some things you can do forever. Given a deep enough shaft, you can fall forever. You can forget forever, and disintegrate forever, and you can laugh for a very long time. But you cannot bleed for long – not you, not citruses, not twites or treepies, not orangequits or plushcaps or jewel-babblers, nor any creature whose vessels flutter with warm, swirling, cell-bearing plasma. Either your leak will mend or you will become void.

Only love can bleed forever; only love has endless blood. Only love's slender drooping tassels can bleed yet grow stronger, bleed yet grow brighter; redder, redder, never spent, never phantasmal-grey. Maybe, if it only gets kicked, then love is love-lies-dented, and in a few days it replumps. But when it suffers a terrible wound, love seems able neither to heal – to grow substitute tissue over its damage – nor to run dry. The other name for love-

74

lies-bleeding is *Amaranthus*, and *Amaranthus* means eternal and unfading, and even after the *Amaranthus* plant fails and drops to the earth, its flowers are brilliantest, bleedingest, sheddingest red.

Love-Bind

Love-bind is an earth-hog. A rampant plant, it stampedes ten metres per season, sprinkling thousands of seeds per year. It binds the forest, wraps and winds it in itself. Forest, are you there? All we see are forms — linden forms, lentisco forms — all engulfed in uncontrollable green love. After years, O trees, of slow-layering labour, of soft blond sapwood maturing into dark sturdy heartwood — you were overcome in one season! Once distinct, once fingertipped with intricate twigs and buds, you are now merely the prop for scrambling love.

Dear people, track your offwandering donkeys with care, lest you get blitzed by rushing love in the green-mad forest and become another of its shapes; for love-bind overtakes all but the rapidest donkey trackers. Moss and mushroom communities are defenceless against love-bind, being stay-put.

Could we marshal all our myrmidons and send them in, with spikes and pikes and poison, to argle-bargle with the love-bind, even they might become shapes: myrmidon-shapes, blanketed in greenleaf. Burly armoured warriors are no safer from being waylaid by love-bind than the slight nymph Dryope was from being transformed into a trembly poplar, no safer than small bland beach animals are from being overswept by seawaves. *Clematis vitalba* never goes feeble-brown; it ever reaches, ever enwraps, like the effusion from some fanatical knitter's speed-needles.

How different the practical plants, like drupelets and teff! For them we have to till, for them we have to hoe, pulling weeds out, pushing pods in, beseeching the sky; then *if* the sky is benevolent our nurslings *might* shuffle thinly out of their seeds. If you have time, you can do a simple experiment in comparative botany: sit down in your yard and chronicle the order in which different plants travel to you. After maybe a decade radishes will creep from the east, with radish ministers. Radishes are like aristocrats, they require an entourage. Then three years later you may notice some cucumbers at your feet, escorted by weeders and slug-shooers and attendants holding humidistats. But earlier, much much earlier, weedy love-bind will have sped to you, surrounded you, leafed you, and kept going! Weedy rambling love arrives within weeks, without squadrons of nurserymen, in an onslaught, its own sponsor!

And if the wind could blow so far, perhaps the moon too would be draped in love-bind, since palest things on Earth, most moonlike things – drudges, and graves of drudges – themselves become swathed once in love-bind's vicinity. Then along with the other shapes you see drudge-shaped love and grave-shaped love. The stars, even, might get dragged into love's all-draping project, the sharp pointy sparkles of Acrux and Spica buffered with leaf-weave – Dryope-stars – if only the seed-bearing winds weren't altogether impounded on the Earth, or if the stars weren't so extra-atmospheric.

Love-bind turns everything into a dubitable green figure. Once-unambiguous mangadous, once-dapper pineapple-gadrooned credenzas, once-natty hatstands – all rendered approximate by overrushing love. The love-trammelled are no longer spruce, no longer chiseled, no longer emphatic. Considering a mystical green ectomorph, one might wonder: 'Tetzel the tavernboy? Beanstan the barrelstaver? Wictred the twit? Maybe a butternut seedling?' Overgreening love converts the Nest to a Was-a-nest, the Wiliwili to a Was-a-wiliwili, Upholsterers to Were-upholsterers. Everything becomes erstwhile; and in this way love-bind is like time and obliteration.

Yet perhaps the erstwhiles are not so desperate to get free. What looks like a tedious costume from without may feel different from within. Perhaps love-bind not only wraps but procures, even translates. Perhaps its underpinnings are not thinking, 'Quit me, Love, I'm no girder', but 'Green me Love, green me on, quit me not, twine me with your fresh pressing weight. Once a success, now a success-weed, I look just like the idiot-weeds over there. Still I wish to never be rid of this levelling leafage, for once my heart was pasty, but now my heart is green, growing-green.'

There are many ways to be transfixed, and no season is safe. If it is winter you may be transfixed by ice; if it is springtime, by fire-finch music or phoebes singing or the squeaky compositions of fox kits. And if it is summer, you may be transfixed, like Dryope, leaf by leaf, by clambery vine-winding love-bind. For love, onslaught-love, beleafs all things.

Please Do Not Yell
at the Sea Cucumber

One nice thing about having bones is that you don't get rerouted every time you run into something. Your route can be quite influenced by your self. Having bones also makes it possible to organise your noise, as in tapping and talking, instead of just swishing in the breeze.

However, when you have tapping bones and talking bones, you are *expected* to tap and talk, instead of just swishing. Each bone comes with responsibilities. Also, being self-influenced can be socially stressful: because they could be proceeding south or east, animals with bones are often asked to answer for why they are proceeding north. Imagine the extra answerability of peewits, who can also go up.

Jellyfish, on the other hand, are less accountable for wherever they are. Even great paroxysms of responsibility have little effect when you are made of mucus. Jellyfish do pulse their bells, but this pulsing influence is minor compared to the influence of the ocean. For instance, the by-the-wind sailor jellyfish is born in the middle of the Pacific Ocean either with its sail tilting to the right or its sail tilting to the left. All the right-sailed ones blow to California, and all the left-sailed jellyfish go left.

When a jellyfish does manage to pulse itself someplace, it is often up, or down – like the upside-down jellyfish which is born tiny and free-swimming, but then, once it grows to be two centimetres across, turns upside down and pulses directly to the bottom of the sea and sticks there permanently, with its feeding tentacles floating up like seaweed. The only way for a jellyfish not to drift is to plant itself. Down there, stopped, the upside-down jellyfish is as independent from the currents as a jellyfish possibly can be.

Most other jellyfish never stop wafting. Usually, dead jellyfish on the sand have simply wafted out of the water. What happens to jellyfish out of water is similar to what happens to bridesmaid hairdos in water. Jellyfish in the water look like pink and green flower hats and bright dripping egg yolks and manes of lions, but out on the beach sand they look like melting plastic bags.

However, plastic bags can also be transformed by

their location: in the water they become handsome like the moon, so that sea turtles mistake them for moon jellyfish and eat them. Eating plastic bags instead of jellyfish saves the turtles' eyes from swelling shut on account of tentacle sting. But plastic causes terrible eater's indigestion and generally kills the mistaken turtles.

Sometimes, while moon jellyfish are being impersonated by plastic bags, they themselves go to Japan to be in salads (jellyfish might seem like they would taste eerie, but in fact they taste like blue rubber bands), or they transfer to aquariums. And once, twenty-five hundred moon jellyfish were selected to join the moon in space. They spent nine days in plastic bags in space and were sent there by NASA because they are able to sense up and down.

Their gravity receptors, also called statocysts, turn out to be quite useful in the water as well as in space, for the life cycle of the jellyfish makes extensive use of gravity. Jellyfish are born midwater, in autumn, when floating eggs and floating sperm float together and form polyps. Then the polyps drop down and live as little necks attached to the bottom of the sea. Like the upside-down jellyfish, a polyp just plops statically in one spot. However, for mobile animals, the sea bottom, because grippable, allows for a lot of dramatic action not always possible up in the ungrippable water

— water which may at any point fluctuate you and strew you and disperse you from your adversaries.

The floor of the sea is where the purple, orange, and red bat stars live, who constantly fight with each other in slow motion — belligerence knows no tempo. On the seafloor, too, the sunflower star steps along on its fifteen thousand tube feet, stepping after the purple sea urchin and the red sea urchin in order to eat them. The red sea urchin usually lasts longer. One way to put off being eaten by the sunflower star is to be eating its feet as it steps after you. Sadly the unfortunate jellyfish polyps have no way to delay being eaten by the dreaded biscuit starfish.

The floor of the sea is also the setting for the potentially dramatic life of the sea cucumber. The cucumbers do not flip or flash or whistle or ever translate into frillier forms of themselves. Neither do they do much mixing with other animals — and mixing usually provides one's chief opportunities for drama. No the sea cucumber eats dirt and its drama is solitary and ultimate. Every year, for three weeks, it melts down its respiratory and circulatory systems and then rebuilds itself. The danger is that if it gets warm or stressed during this restoration period the poor frail cucumber will burst, expelling all its softened heart-soup. Please do not yell at the sea cucumbers.

In the springtime the jellyfish polyps develop little

decorations on top of their necks that turn out to be medusa-buds, medusas that will go floating up, away from the floor, away from the fussing bat stars and exploding cucumbers, to drift around in the seawater with peaceful sea gooseberries and stable sea walnuts.

The medusa form of the jellyfish is the famous one, more famous than its stationary polyp form. However, there are some polyps which never actually sprout medusas, but instead assemble their small basic plain selves together into a medusa lookalike. The man-of-war, for example, appears to be one individual, like Leo Tolstoy; but it is actually many individuals living together as a colony, like Leo Tolstoy. The man-of-war is a joint project – many little polyps joined up – so some of the polyps swim for the project and some reproduce for it and some eat for it and some serve as dangling tentacles. The siphonophore, a twinkling blue rope longer than a whale, is also such a colonial project.

As for their twinkling, many jellyfish will only twinkle when jabbed. They do not spontaneously light up. Nevertheless, the jellyfish's green fluorescent genes have been successfully transferred to piglets who then glow whether they are jabbed or not. Fluorescence and the option to stop fluorescing are not always concomitant; the new piglets have permanently fluorescing snouts and hooves to ruffle up truffles with.

But while the jellyfish is able to beam or not beam at will, there are other things that it cannot stop doing, like poisoning. Jellyfish have to be so poisonous because they are so delicate — too delicate to struggle with their food. If their food thrashed about, it would shred the jellyfishes' curly gauzy watercolour streamers: therefore they must be noxious enough to subdue their food at once.

Trailing long death-streamers beneath you works well when you have no teeth and are not good at steering; however, death-streamers, unlike teeth, do not allow you to target your victims and hence you will sting everything you touch night and day — both the things you can eat and the things you cannot eat, like benign swimming persons, who, when stung, may experience 'a feeling of impending doom' and then sink too quickly to be eaten. Jellyfish venom is similar to scorpion venom and to funnel-web spider venom, neither of which is addictive.

Now there is a little silvery butterfish, which, because it withstands the ever-sting of jellyfish, reveals them to be in some circumstances gentle, even henlike: the butterfish spends the first months of its life tenting in the soft poisoned tentacle-curtains of a jellyfish. The silvery butterfish is probably poison-resistant, since it eats as well as inhabits the jellyfish tentacles. (How to eat something larger than yourself:

bite off chunks.) A decent question here is whether the host jellyfish is really hospitable or if it is just insensate. After all, sometimes you do not notice when fish are swimming through your beard but that does not mean you are so hospitable.

What *do* jellyfish actually sense? Because they light up when jabbed, they probably sense jabbing. And they must sense deepness, as demonstrated by the hulaskirt siphonophore, who regulates its deepness by augustly changing the amount of gas in its float: if you can regulate something, surely you can sense it. Also, if you swim away from something, you can probably sense it, and jellyfish swim away when submarine rovers shine their spotlights on them. Sometimes they even melt in the spotlight: melting is a good sign of detection.

Another reason to suppose that jellyfish sense light is that they live *in* light, like tomato frogs and bears and grass. Even grass senses light, although slowly, while light is meaningless for tapeworms. Light sponsors its own comprehension.

Another clue that jellyfish sense light is that they *emit* light. Consciousness, of course, sometimes falls short of emanation, and jellyfish could be like flowers and people, who express more than they know. But jellyfish seem to emit light deliberately — sometimes a jellyfish will even drop a glowing tentacle onto a predator and then turn its remaining lights off, which shifts everyone's attention to the predator, suddenly conspicuous with a fantastic flashy tail.

In the end, though, the best evidence that jellyfish can sense light is that they have eyes. Why do jellyfish need eyes? Bees, of course, need eyes to see other bees dancing; birds need eyes to watch the stars; smart bombs need eyes so they can stare at their targets; and boa constrictors need their infrared vision to sense tiny hot animals in the forest. Scorpions and turtles may or may not need eyes, as scorpions have from zero to twelve eyes and turtles sometimes travel with their eyes closed.

Jellyfish, however, do not need to communicate about faraway nectar troves, nor do they navigate by the stars, like birds. But still they have eyes.

Most animals have either keen eyes *or* sensitive eyes: cats have iridescent tapeta in their eyes for gathering the palest traces of light; but all that gathered scattery light in their eyes, then, prevents cats from perceiving fine details. And hawks detect details, but since they do not have tapeta for collecting flickers, they must depend on the sun to boom down obvious light for them to see by. Your blessing is your curse and your curse is your blessing. Because you see details, you cannot see hints of light; because you see hints of light, you cannot see details. You would need diverse eyes if you wished to be equally penetrating and sensitive.

You would need to have eyes like the box jellyfish, with its sixteen light-sensitive eyes and eight acute cameralike eyes — all twenty-four eyes hanging down on stalks.

However, you would also need a brain.

But maybe that is not possible; maybe, in fact, the brainlessness of the box jellyfish is a direct consequence of its tremendous powers of sight. Perhaps neither the animal nor the prophet has been invented who could process so thorough a vision. It is disquieting enough to be hyperacute *or* hypersensitive; perhaps being both would very soon melt your brain and leave you quiescent, hanging transparently in the giant dancing green waters of the world.

You Are Going to Fly

Once, a friend and I followed a moth that trudged across a whole grocery store parking lot. It was nighttime — not really a safe time for trudging moths, people generally being too tired, by then, to stop the car and wait. I said to it: 'Moth, why don't you fly? Why do you waste your wings?' But my friend, much better than me at allowing winged things to walk, said that maybe it missed being a caterpillar. We decided this: it is a good thing that the caterpillar stage comes before the moth stage, instead of the other way around; moths, if they like, can take long nostalgic walks across parking lots, while caterpillars could never indulge in long nostalgic flights to Mexico.

'Caterpillar, you are going to fly,' everyone says to caterpillars. 'You will be transformed!' People think of them as protomoths and protobutterflies and are impatient for them to convert into their more extraordinary selves. This is understandable, of course: when that which is like a plodding lozenge turns into that which is like an angel, everything that belongs to the lozenge's time seems mere preliminary. But think of the nervousness for the lozenge! What sort of disposition could bear the pressure of such a drastic and imminent exaltation? Caterpillars, nonetheless, remain calm, eating their tobacco and milkweed, enviably imperturbable in the face of a brilliant future.

Yet they *are* perturbable — if not by the future, then by elements of the present, for they are little and plush-soft, crushable, eatable, drownable, freezable. A few have spines, like the hickory horned devil, or intimidating tufts, like the dagger moth; a few have antifreeze to keep their blood from turning to icy slush, but most do not. Most can only inch away from danger, antifreeze and intimidating tufts notwithstanding. Any danger that cannot overtake a caterpillar is no danger at all, a trifle.

Because they cannot run (for caterpillars have only six real legs — the rest are fake: mere stumps to keep their hind parts from dragging and getting scuffed), caterpillars have to do other things when threatened. Some make themselves unpleasant: black-etched prominent caterpillars send out two foul-smelling pink tentacles from their back end and wave them around. Monarch caterpillars are foul-tasting. (Entomologists use that word 'foul' often when referring to the flavour of a caterpillar. They are rarely more specific than 'foul' or 'tasty'. I expect that is because they are leaving the assessment up to birds, and birds have a very binary approach.)

The azalea caterpillar, a black-and-white plaid caterpillar with a cherry-red head and legs, when disturbed, arches up its head and thrusts it back, like a hairpin, and arches its tail up like an *S*. To be honest, it looks more electrocuted than scary when it does this. The yellow-necked caterpillar twists itself into the same shape, except that the yellow-neck vibrates as well, which *really* brings electric trauma to mind.

Many caterpillars defend themselves not by striking fear in the hearts of their predators, but rather indifference. The large maple spanworm looks like a twig; the viceroy caterpillar looks like a bird dropping. This is not as exciting as looking like an anaconda, but when you are very small, and wingless, one of your main goals in life is *to not be exciting*. And speaking of unexciting — I think it is safe to say that woolly bears have one of the least advanced defence mechanisms among insects, although theirs is the reaction with which I most strongly identify: when distressed, the woolly bear rolls up into a ball.

Their other main goal, besides not being eaten, is to eat; we are lucky we do not have to share our tables with caterpillars. Elephants attest: in six weeks, in Botswana, the mopane worms eat thirteen times as many tons of mopane leaves as elephants do the whole year long! But for such ravening creatures, caterpillars are finicky. The mopane worm prefers mopane leaves, the cloudless sulphur prefers wild senna, the sleepy orange eats senna and clover, and the darling underwing eats willows but prefers black willow. It is not an effortless venture for the Earth to keep caterpillars happy, or to keep them full.

I recently saw the last five minutes of a reality show where lots of women were vying for one man, the women behaving like witchety grubs frantically trying to bury each other in a Bucket-O-Bait, or spiders stinging each other's heads, eating each other's legs. Not having seen the beginning, not knowing the premise of the show, I surmised that some amount of money was in store for the winning woman. (What other incentive would induce them to participate? Unless they were coerced – but it cannot be legal, I thought to myself, to detain so many women and force them to act like insects.)

However, when they showed one of the youngest ones, filmed burbling after the big introduction banquet, I was nonplussed: not a mercenary at all, she seemed genuinely hoping to fall in love. She said, 'The moment our eyes met, I *knew* there was something special between us! And I felt butterflies in my stomach, like I haven't in so *long*!' And she declared, 'I *live* for butterflies!'

I am guessing that this sweet, moony girl later regretted her announcement, which must have induced thousands of entomologists the world over to write her letters: 'Dear Tiffy, We do *too*! We live for butterflies *too*!' Entomologists are notoriously excitable, and notoriously ingenuous when it comes to reality shows. But it is possible that a more guarded entomologist, maybe from the Ukraine, happened to walk into his teenager's room just as this particular interview was airing and directly posted a sensible letter the next morning to

> *Dear Miss Brown,*
>
> *After witnessing your show on television last evening, during which you made the emotional statement regarding butterflies, I feel compelled to write to you and inquire: are you not aware that most butterflies, after they have acquired wings, only live for two weeks or less? That they spend months or even years as eggs and caterpillars and pupae? This is why you haven't felt that fluttery feeling in such a long time, Miss Brown: the life cycle of the butterfly is protracted. If you really do live for butterflies, as you say, and*

provide your thoracic cavity as a butterfly garden, you must
be patient and remember that you are living for their eggs,
as well, and for caterpillars, which require months worth of
nourishment, and for pupae, which can wait for years before
they emerge as butterflies to give you a few days of fluttery-
feeling and then die, or migrate. But with all due respect,
Miss Brown, I suspect that you do not live for butterflies at
all, but for the fluttery feeling they offer. If this is indeed the
case, may I suggest that you let the butterflies out and fetch
some Mexican jumping beans instead?

   Regards,
   Dr. Osip Iwasykiw

But, in the end, whether people know how small
a portion of time is given to the butterfly, how large
a portion to the caterpillar, does not matter. For
they can never infect the caterpillar with their anx-
ious urges to 'Become!' A small apple-green caterpillar who climbs a toadflax plant, who somehow loses a
foothold while walking across a stem to get to a leaf,
slips and is hanging on by only two crochet-hook feet,
the wind swinging it back and forth over the creek, is
not thinking, 'Alack! I shall fall into the icy water! I
shall be swallowed by a fish! I will never, now, wrap
myself in silk and wake up with powdery, iridescent
blue-and-green wings, fly away with them to fields of
cornflower, and mate, and feed on the tears of wild

buffalo! My life, my eating, my climbing — it has all been meaningless!'

Rather, it thinks, 'I'm swinging, I'm swinging, I'm swinging.'

God

The hoopoe and the bat do not say this word. Neither do the eagles or the vultures or the black vultures. The hyena and the wild goat and the night creatures refrain from using it, too. I have found stoats busy in the gutters, doing what they wish, but never uttering this word. A ferret may slip off one's lap, stalk away along the floorboard with a bend in his back, crush under the back door and leave. Outside in the dark, nobody knows for certain what escaped ferrets do, but they've never been heard saying the word, or seen forming it silently with their mouths.

The people say the word repeatedly, and the more they repeat it, the less I can understand it: listening to words I do not understand is like swallowing stones. With each repetition of the word it is like I am given another stone to swallow. I can't keep up, for it is hard to swallow stones, and I get behind. I have stones filling my mouth and stones in my lap, and stones falling out of my pockets, and the stones keep coming heavy and hard.

The word refers to someone no one has ever seen. Perhaps this is why people say it over and over, as if repetition of a word can make up for the absence of its referent. They say it pleases him, to say his name incessantly. They sing it in songs and chant it together and broadcast it loudly on the radio, on signs. Perhaps it pleases him. I do not know. It does not please me.

Some evenings as I sit there with all these stony words piling up on me, I get so overwhelmed that I become indifferent, and I spit the stones out and let the heap on my lap fall to the floor, and I walk away and go out the back door. The escaped ferrets are out there. The hoopoe and the bat are out there, and I listen to them, and I drop into the pond and swim with the black eels, and I listen to the eels. I listen to the jackrabbits and the javelinas and the sandhill cranes, for they are all out there. And so is he to whom the overuttered word refers. He is there because his words are there.

His words do not rain down like rocks on those he speaks to; they mount up with wings or leap through brambles or swim blackly in ponds. They sleep hanging from trees, stomachs full of hunted insects, or grow tall and imperious and leafy in the forest. Many, if not most, of his words hope never to be heard — rooting blindly through their dirt-homes or proliferating on the tops of mountains, they are dismayed when they are discovered, and rush away. His words are not repetitive: the only thing his words have in common with each other is that they are strange and they are themselves — they move on their own, through gutters and caves and swamps and the sky, and some of his words, when they get tired of hearing his name over and over, and wish to hear him speak, escape out the back door, like ferrets, like me.

MEMORANDUM
TO THE
ANIMALS

And of every living thing of all flesh, two of every sort shalt thou
bring into the ark, to keep them alive with thee.

—GENESIS 6:19

Unfortunately, Animals, we are not going to be able to bring all of you with us this time. Last time there were eight humans on board and at least two of each of you; but that was a sentimental era and God was a sentimental fellow, like the old pack rat up the road who won't give up any of his whim-whams. Bringing two of every sort of creature onto the ark meant bringing all seventeen hundred species of scorpions, and all the blotchy toads and testy wasps and malevolent snakes and countless other creatures as unnecessary as insanity. This time around we are in charge: producing our own cataclysm, designing our own boat, making our own guest list, which does not include Every Living Thing.

That first, ancient boat we have retrospectively christened the *Fantasy*; today we sail in a boat called

Reality. Realistically, logistically, it would be too complex to try and save every single kind of you multitudinous, miscellaneous creatures. Dormitory assignments would be a nightmare, because we don't know who might eat whom or who might die of social stress. We'd have to go nectar collecting beforehand for sustenance for the sugarbirds and gather angelica blossoms for the hoverflies, and if little blue poison frogs were going to maintain their poison, we'd have to bring along extra oribatid mites for them to ingest, and special leaf litter for the oribatid mites to ingest.

Anyway, we need the space for *our* works and wonders. Many of you are being superannuated because we must give priority to our machinery: our televisions and computers and refrigerators and cars, trucks, airplanes, combination microwave/convection ovens with auto-timezone adjusters. We will still bring a few of you with us, especially those of you with rumps and ribs (please refer to the Keep-Alive List). But we are not going to waste time hallooing for the bushbabies, waiting for the mayflies to drift in and the kiwis to materialise. We are certainly not going to stand around until the tortoises figure out what's going on.

If you are concerned about the devastation of your genetic type, and you do not see your name on the Keep-Alive List, you might think about clumping some vegetation together into rafts on which to rescue yourselves. You animals (slugs, bats) who cannot assemble your own rafts, or whose spindly legs (dromedaries, moose) are liable to poke through a grassy vessel, or who are graminivorous and oblivious (sheep) and who would nibble through the rescue raft: know that the extinction of your type is not necessarily the extinction of your glory. You can live on in the imagination, like the angels – although like the angels, you are likely to be simplified.

To *properly* live on in the imagination you should have someone who really knows you, who knows the pitch of your buzz, the rufous hue of your throat, your fondness for comfrey – who has watched two of you golden frogs, separated by a loudly gushing stream, waving your hands in semaphore fashion. As the Holocene is winding down, don't be foolish like the okapis, who use their acute hearing to detect and avoid human beings; they have sabotaged their chance to survive in anyone's mind.

But even the most knowledgeable imaginer is a fallible vessel – like an iceberg, liable to get carried away, to melt, to be at bottom very very blue. Don't you know, Animals, nothing lasts forever. The Holocene was the Age of Miscellany, the Age of Pandemonium, the Age of which Noah's Ark was a microcosm: of thirty million passengers, only eight were human.

It was a noisy, dirty, dangerous, eccentric, anarchic, inefficient, rowdy ride. The copassengers knocked us over a lot and sometimes we fell over laughing. But now we are proceeding into the Age of Efficiency, the Age of Sanity, the Age of Refinement. We salute you, Animals, as we salute the unruly Holocene; but the future belongs to us.

II.

THINGS

OF

HEAVEN

The Same Old Joy

Every Monday at noon, in Fry County, the Emergency Warning Sirens are activated for a drill. They produce one high, unwavering loud note for exactly sixty seconds, after which they perform the All Clear tone. Then they are inactive until the next Monday but must remain at their stations: if there is an accumulation of snow over four inches, parking regulations go into effect and the sirens are to sound at 7.15 a.m. and 12.15 p.m. Sometimes, during hailstorms or lightning or strong winds, the sirens are used to disseminate Weather Awareness.

They are installed on top of sixty-five-foot poles around the county, and they rotate to cover an acoustic radius of 1.86 miles. The Fry County Emergency Management Agency has sixty-two Emergency Warning Sirens, with four new sirens scheduled to arrive in the next year, to provide better coverage. The Monday tests verify that they are in good working condition, but once every five years or so a siren malfunctions, chanting and trilling exuberantly at 4.30 in the morning. This alarms many residents, who call the police. Do I need to move my car to the other side of the street right *now*? Is there a threat of grass fires? The police assure them there are no emergencies and send someone to silence the offending siren.

Other than the occasional glitch like that, sirens are very effective as notification devices, for they possess extraordinary powers of projection, cultivated of old: from the flowery island of Anthemoessa they needed to be heard, across miles of sloshing waves, by distantly sailing ships. The sirens could sing the fish out of the water and the stars out of their constellations; their song was that loud and sublime and unsettling. But they were not very shrewd musicians, in terms of their career. Astute musicians will steadily cultivate a following: if ten people come to the first show, and they like it, and survive it, thirteen people might come to the next show, and eighteen people to the next, with friends bringing friends until a loyal and constant fan base is established. One's performances should be groovy but benign so the audience will live on to recruit more audience.

But capacity is hard to suppress: if a cat can jump

from the curtain rod to the top of a person's head, she certainly will. Contemplating the sandy bones of their latest audience, the sirens would vow to sing less devastatingly next time – on their own they practised thready versions of 'Lavender's green, diddle diddle, lavender's blue'. But as soon as they sensed someone sailing by, their vows evaporated: they would start feeling ecstatic; their voices would swell, and deepen, and soar, and then it was all over. They sang deliriously, mercilessly, driving the hearers wild, drawing their haunted hearts into the sea. By the time the song was finished, the sailors had all flung themselves overboard.

There are gifts, like the ability to do cartwheels or make strudel, which are clearly edifying; but another kind of talent results in universal deprivation. Who hears the song is deprived of his life; who sings the song is deprived of her hearers; and the ships all toss around sailorless. For thousands of years there seemed to be no way out of this ridiculous cycle of song and destruction. However, as demonstrated by their Fry County duties, the sirens have apparently found a constructive channel for their talents. Some of the most responsible ones even get mounted on top of airport security vehicles, which drive around and scare little black birds off of the runways, to prevent bird-plane collisions.

Does it bother them to have to sing such minimalist music? They who used to live on a warm sandy beach under green palms, eating sweet flowers drenched in coconut milk — is it boring for them to perch on small metal platforms on top of tall poles, waiting seven days until they can finally emit one note for one minute? Do the sirens pray fervently to the god of weather and atmospheric phenomena to send tornadoes, windstorms, disasters of any sort, to have more excuses to sing — effectively subverting their function in emergency management?

Or have they, strangely, found an audience that compensates for the monotony of their present employment? Of course it is no longer *people* who respond to the sirens, except to repark their cars or huddle in hall closets; neither is it frogs. Although frogs fling themselves into muddy puddles upon hearing the sirens, they do that anyway and it is not the effect of frenzy. But there *are* still hearts inflamed by the siren song, souls that still contract a delirious, inexorable passion upon hearing the emergency management employees. Every Monday, exactly at noon, all of the beagles in Fry County lose their minds, are transformed into loony fools, bursting into wild and painfully sublime songs of their own. They sound like they would leap into the sea if there were a sea, instead of just a yard.

Sirens used to sing a song people would follow to the other side of life. They were free like thunder, and dangerous like tornadoes, and enchanting like fire. But for all the drudgery, the artistic restriction, the tedious utilitarianism of municipal appointments, perhaps when the sun comes up on Monday, and the sky turns golden blue, the Emergency Warning Sirens close their eyes and sigh, and hug themselves, and hum under their breath, swaying back and forth, high upon their stations, swaying with the same old joy, anticipating the noontime drill. One note, per week, followed by the All Clear tone, is enough, as long as your district still contains some hauntable hearts.

The Safari

When the past happened, it was as strange as the present, as anarchic and wild. Events stormed out of nowhere like obstreperous hippos. Once the past is over, though, you begin to administrate it, locking the days in cages and assembling them by genus and writing explanatory plaques for each. Plaster habitats with synthetic plants and painted savannas showcase the inhabitants, and maps suggest a reasonable, instructive sequence for a visit. To make sure none of your memories attacks or defects, you place them behind deep moats and electric-shock cables and glass barriers – whereby, also, no memories may infiltrate, for not all of them are selected to be in the park. You build snack stations and for special holidays hire singing Victorian mannequins. Thus does the insane jungle-past become a nice zoo park. Thus is the pandemonium of the past civilised, accredited, receiving a retrospective order and sidewalks. Thus is it made secure and navigable, and fun to visit; and thus is it easy to exit.

But sometimes, as you stroll confidently by the cages, you notice that the internees stop examining their toes and turn their backs to you: fuzzy grey with lavender rumps, or scruffy black with long tufted tails. Or they commence to making dung. Or they pace contracted figure eights in the corner behind the imitation log. Or they don't even blink when you appear; they just lie on their plaster gradients staring at the fluorescent white gridded ceiling, while the bonneted diorama sings, 'Ding dong merrily on high! In heav'n the bells are ringing!'

Then you wonder if these exhibits, so perfectly controlled, are sincere. The memories seem diminished into listless props for their own interpretative panels. Hadn't some of them, eyes afire, once chased you down and pounced on you, growling in your ears like berserkers? How are they now so quashed and torpid? Is it your rectangular management of them, your marshalling of the desperadoes into a testament of purpose and causality? Is it your group tours for luncheon ladies? Is it the singing ding-dongs that so dispirit the inmates?

You wonder what it would be like if your memories were to go afield again, unpoliced, set free from their subjugation to Biography, free once more to haunt wheresoever they would. You impetuously take your skeleton key and run through the zoo unlocking the cages, throwing open the main entrance, liberating the captives to hie themselves to the greenness of groves and the concealment of burrows and to the air, if they are winged. And then you follow them out, putting yourself at the mercy of memory.

The first thing you see on your safari is nothing. With the abolition of their coercive exhibits, your memories disexhibited themselves, heading deep into a screen of fog and magic guarri trees. At least in the zoo, they were on display, even if they all were nodding off. But then you notice two tall thorn trees, far away, shuddering. One tree bends forward as if it is being eaten, and then two soft horns rise above the other tree, and dark eyes with long eyelashes look. Elusive, quiet memories, like giraffes, may or may not raise their soft plush horns above the trees. If they do not, all you can know is that treetops tremble quietly in the mind.

In the zoo of your memories you had allowed only plump and thriving rodents, but now you spot, underneath a log, a wretched creature, striped brown and cream. Like all mice this mouse was once shiny, fat, and admirable. With time, however, it began to unravel, growing sickly and despicable. Now it tarries dishevelled under a log, like a regret, staggering out for a seed sometimes; but it does not die, for mice and memories do not like to die, even if they are miserably changed.

For a while, after the mouse, you feel like you are on safari in a hennery; you see nothing but hens, hens, little blue intent hens hurrying past you; hosts of little pedestrian hens pressing by. They pay you no mind and hop over the holes, where naked mole rats dug down into the soil, never to be seen again, never to suffer from the brutal sunshine and brutal air on their fragile pink skin. Some memories are so fragile they bury themselves. Many are very reticent and many are pedestrian and time has dilapidated some.

With your expectations accordingly muted, you walk across a grassy golden plain and then pass over a hill to a pond. Storks and egrets and cranes are fishing there. Upon seeing you, one slim white egret flies up. This startles all of the birds, white and brown and black and yellow, into the air. They follow one another up in steep circles, descend to land with hanging legs, then fly up again in a slow, jubilant throng. Without cages to keep them apart the days are hard to study, for as soon as you surprise one white egret, it surprises the slaty egrets and the cattle egrets and the yellow-billed egrets and the dwarf bitterns and the squacco herons and the woolly-necked storks and the flamingos and the spoonbills, and then the sky is motley.

After this you look for a memory that won't turn into a crowd, a solitary one, so you stop and peer into every jackalberry tree, looking for the long rosetted tail and the heavy round paws hanging down from a transverse branch. But as deeply as you peer into tree after leafy tree, you do not see her; neither can you hear her growling in her dreams. You thought she would always be immediate, this dear and dangerous animal, but by no intensity of wishing are you able to summon her. How obstinate are memories in trees, yet how dull and groggy do rooms render them!

Then you cross a stream, and as soon as you step one boot into the water four little moss-green turtles swim up to you; then hundreds of other turtles in the neighbouring pool do too. They do not pinch you with their beaks; they just scramble up to you, wanton, accosting your shins, as if to say: remember us, remember turtles; as if they wanted you to take them all with you, for you to carry hundreds of turtles in your arms.

Not all forgotten things are frantic to address themselves to you. Even as you are being mobbed by little turtles, you see three grey forms running away. They are grey like rocks and have lithe legs and twisty horns, but you do not know what they are. None of your memories took their explanatory plaques with them when they were released; some of them run farther and farther away from apprehension until they are hiding in aardvark holes.

Maybe some people's minds are like libraries, with memories like books, and even if the exits were left open the memories would still be sitting on their shelves in alphabetical submission. Maybe some people's memories are like furniture, useful: chairs to sit in and desks to work at and urns to bequeath; and maybe among many normal memories, some people have an unsnuffable flame, which every several years gets aggravated and burns the whole mind down. You happen to have one of those minds inhabited by memories like wild animals, with wandering ways of their own: diffident giraffes, changeable mice, milling birds, clamouring turtles, a few that harry you and many multitudes that skirt you.

If you were to live long enough among your emancipated memories, you would learn to see their manifestations, to know from the trees broken in the mud which elephant had trod there, or from the silvery trail across the rock what snail had toiled by. You would learn the sensible distance from which to meditate on crocodiles, lest they pivot and unleg you. You would understand, from acute and recurrent consideration of your memories, something of their nature. But the longer you live the more memories you acquire, which means understanding will always lag behind experience. And since the memories influence each other, each new arrival means more intricacy of interactions: some memories genially bite each other's tails and some bloody each other and some chase others off the deep end.

Alone in the open land of your wild and shifting days, you follow huge round three-toed footprints through the scrub until you see something standing far away. It has one horn so huge it looks like it should tip forward. It stands still for so long; why does it stand so still? Is it reluctant like the giraffes? Or in disrepair like the mouse? It does not look sick, but the ruin of a rhinoceros may long happen subcutaneously, while for mice, fur is an early manifester of fallibility. Then the wind kicks up and the rhinoceros gets a whiff of you and, like all rhinos with whiffs, takes off running in the direction of its horn — toward you, as it happens. Hence what seemed a weird but benign memory from a distance, while you sat behind your binoculars and formulated plausible theories about it, may all of a sudden charge and leave you, the rememberer, in shreds.

Now you would like to go, to limp away, to leave the premises; it was a reckless plan, to adventure among these temperamental entities, with no rifle and no experts and no truck. But how can you leave when there is no perimeter? You have freed all the constituents of your whole lifetime and they have distributed themselves over the entire area of your mind to ruffle through the dew and root through the soil and lurk in ponds. There are trumpeting ones that could flip and trample you, or wrap their trunks around your wrist and dangle you in the air; fanged ones, demented ones, small vulnerable ones like klipspringer lambs hidden in the grass, others leaping on every side, sleeping under shrubs, huddling with their tails twisted together.

Perhaps it was smartest, after all, to collar your memories and isolate them, sedating the irascible ones, banishing the grotesques, systematising the rest; maybe coaxing a lion into a wheeled cage on occasion and pulling it eminently around town for the neighbours to see. Maybe it was best to let only the *shadows* of your impounded memories touch you; shadows usually being safer than their begetters, as for example axes and icicles and porcupines. But it had begun to be so mind-numbing, to see your memories in their state of moldering stupefaction; even the spider wasps were insipid, in their stuffy glass boxes. How could you not *unloose* them? Wouldn't it be tempting to liberate even the Zoological Park of Ghouls, just to see them grisly again? Enervation is a most unbearable sight, even in ghouls and hellhounds! Anyway, the emancipation is irreversible – will a warthog, once freed, consent to being recaptured?

It is now getting to be late in the afternoon. The bushwillows are backlit by the sun, their leaves young and electric green-yellow, vibrating in the breeze. A herd of zebras with stout round hindquarters and stiff Mohawk-manes comes romping through the willows and runs in restive circles around you. With their streaks and spiky manes they seem electrified too, and the ground throbs with their energy. They pound around you, round and round in the fledgling glow, and transfix you with their black and white, black and white. The skinny twitchy foal freezes, then scurries after its stocky thudding mother, then freezes again, doubtful. It seems strange for memories to run out of nowhere and circumgallop you like this. It seems strange for there to be memories; strange for there not to be memories; strange for time and space to have dissevered you from anything so radiant and pounding; strange for your mind now to disregard time and space so completely.

Before you have spotted even a fraction of your days, the sun will be departing. The antelopes might be anthills. You might feel some passive languor as sight becomes impossible; and some relief, too. But right before the sun goes down, the light will get more intense, and your memories will rouse and trumpet and stamp and bawl and honk and become whiter-teethed, greener-feathered, browner-furred, sharper-horned, softer and more violent, as if time's filmy layers had dissipated and your entire past were happening anew — the past as new as the present. Your memories will sting and stain your eyes as this final sunlight is translated into red-gold lions and gilt-orange floating birds with long ribbony tails and yawning wet stump-toothed hippos. They will cast quills down your throat and wrap you round with their claws and surge your ears with bellows and send your blood rushing and charging with the extremity of their presence.

And then one of your little days, like a kingfisher, will fly over the waters, diving down beneath the opaque golden surface of your mind, where swim your earliest, submarine memories. What is caught is a tiny primeval memory that should mean nothing, a throwaway. Yet when pulled out of the water, gripped in a birdbeak, lashing the air and throwing flashing grapefruit-coloured waterdrops from its glittering, tiny perishing silver self, this forgotten, underwater matter will suddenly mean all the world to you — the long lost glittering hour that means more than age, more than logic, more than lore.

Then the sun and its light will truly leave. You will receive the weight of darkness in a hundred hushed blows, each heavier than the last. There will no longer be any way to apprehend your days; the stars will come out but they will only illuminate themselves. They will not send enough light for you to see your still-stalking, still-flying, still-ferocious, still-nearby memories. Sometimes we do not get the type of charity we want from the stars, and then we fault them for being selfish. But what? Would we rather that no thoughtless stars lighted themselves — that our days escaped us in grimmest gloom? Is beauty not a form of philanthropy; and are the stars not the most beautiful fireflakes? Will the stars, at length, not grant us a firmament of shivery, glinting, white-gold flames under which to relinquish our days?

Stairs

A long time ago I saw the tops of the tallest trees from a window. The little octagonal windows here are set high, three feet above eye level, so that you can look up at birds and snowfall but not down on grass or ice floes or bonfires. But I have not seen a bird for many days; it is just blue out. Stars are my bonfires, blue is my diaphanous land.

How nice it would be to see proof of wind out the windows — a dust devil, or some fluffs of cottonwood blowing. The stars are too firm for wind. I miss the exquisite manifestations of wind and gravity, even if I do not miss being subject to them. Sometimes I think that, to see a waterfall, I would fall, like Alice; I would dive backward and twist down all those stairs, except that my ascent has acquired the helpless nature of a descent: I am Alice ascending.

Gravity has started to lose much of its irresistibility at this altitude. People who are afraid of heights would do well to remember that if you just climb high *enough*, you can reach a point where you are no longer in danger of falling. If I tried a swan dive now, there would be nothing drastic about it — I would make a dawdling arc and lightly land on my hands, two steps down. It is like living on a very small planet, which has just enough mass to keep its inhabitants on the ground, but only if they want. Very small planets are sweetly resistible.

This planet is neither resistible nor sweet. In fact it held me flat to it all my life before I started up the stairs. I was pale thin waterweed, stuck to a massy planet.

So when I discovered the tower with the small gold plaque reading 'UP' in the woods behind our house, I said, 'This is for me.' Inside the tower, the stairs are covered with a thick crimson carpet and there is gold wallpaper patterned with soft ivory fleurs-de-lis. The banisters are dark and gleaming in the light of the oil lamps set in the walls. It smells hushed and musty, and every few hours I come to a landing with an imperious-looking chair with carved lion's paws for feet. At first I stopped every seven or eight steps; now I hardly stop for the chairs.

And very soon I will be able to float up to a window, and then I will be able to look down at the waxy blue earth with its waxy white clouds, and then I will be able to rise the effortless rest of the way to the top of the stairwell. I will miss the freshness of grass crushed under my feet, I will miss the wasp-sting, I will miss seeing the pale green praying mantis sway and hesitate and look around before jumping into the air and flying away. But who . . . *who* . . . does not miss everything?

Sail On,
My Little Honeybee

There is an altitude above every planet where a moon can orbit forevermore. In millions of miles of ups and downs, there is one narrow passageway of permanence. If a moon can reach this groove, it will never crash down like masonry nor drift away like a mood; it will be inalienable; it will circle its planet at the exact speed that the planet rotates, always over one site, like the badlands or Brazzaville or the Great Red Spot, so that the planet neither drags the moon faster nor slows it down. Moons not locked into this synchronous orbit are either being perturbed up or down.

The law is stringent about this; there are no clauses; and all moons are dutiful followers of the law. But, as all good followers of the law discover in the end, unless you happen to roll onto a track precisely determined by your mass and your planet's mass, the law ejects you or dashes you down. One moon in our solar system has achieved synchronous orbit, being pledged forever to its planet — Pluto's moon Charon. The other 168 moons have not.

Mars has two small moons whose names mean 'panic' and 'terror'. Phobos looks like a potato that experienced one terrible, and many average, concussions. Phobos hurtles around Mars every eight hours, which is three times faster than Mars rotates, which means Mars pulls it back and slows it down. Slowing down makes a moon lose height; in the end Phobos will smite its planet, or else get wrenched apart by gravity into a dusty ring of aftermath. Mars's other moon Deimos is a slow and outer moon — an outer and outer moon — someday it will be a scrap moon, rattling around in the outer darkness, where drift superannuated spacecraft and exhausted starlets.

So fast moons slow down and slow moons speed up, and only during excerpts of time do planetary dalliances appear permanent. Our moon through many excerpts — the Moon — is a slow moon. Thus it is speeding up, thus it is falling up, coming off like a wheel, at one and a half inches per year. Let us now reflect upon the Moon; for the Moon has long reflected upon us.

To get an idea of the relationship between the

Earth and the Moon and the Sun, find two friends and have the self-conscious one with lots of atmosphere be the Earth and the coercive one be the Sun. And you be the Moon, if you are periodically luminous and sometimes unobservable and your inner life has petered out. Then find a large field and take three steps from the Earth, and have the Sun go a quarter mile away.

For an idea of how long your light takes to reach Earth, sing one line from a song, such as 'Sail on, my little honeybee' and that is how long moonlight takes. The Earth can sing the same line back to you, to represent earthlight. 'Sail on, my little honeybee.' As for the Sun, he should sing as lustily as sunlight; have him discharge the song 'I Gave Her Cakes and I Gave Her Ale', which is eight minutes long, which is how long sunlight takes to reach the Earth. Also the Earth may sing to the Sun and the Sun to the Moon and the Moon to the Sun, songs of representative length.

Now keep singing and everybody spin and the smaller two of you orbit the next largest rotundity. Now as you, the Moon, go around the Earth, do not circle perfectly, as if you were a mill horse, or an idea. You are not an idea; you make the Earth's heavy blue waters heave up and down! Circle asymmetrically, then, like a small coplanet; truly you and the Earth *both* orbit the centre of your combined mass, called the barycentre. Of course, if you and the Earth were equal in bulk, the barycentre would lie exactly between you; you and the Earth would pass your lives in social equilibrium, like the rooster and the pig on the carousel. However, as the Earth is eighty-one times more massive than the Moon, the barycentre is eighty-one times closer to the Earth: thus the barycentre is *inside* the Earth, though not at its centre. This means that the Earth orbits a point inside itself. The Earth is a self-revolver, nodding slightly to the swooping Moon.

Now the Earth does not *look* eighty-one times as massive as the Moon — in fact it is just four times as wide. To address this perceptual difficulty we will interrupt our lunar reenactment, and consult philosophy. Let us refer to our index of philosophies and select one known as Interiorism, which says that truth is to be known by introspection. To discover why the Earth acts so central and the Moon so obsequious, let us not measure yards but consider inward differences. The Earth is not gigantic and the Moon is not slight, but the Earth has a core and the Moon does not. Or rather, if the Moon has a core, it is undetectably small and inert, like a frozen mouse.

How do we know that the Moon has a mousy core? Whoever really has been a Lunar Interiorist? Here we shall invent a philosophy and call it Imaginative Exteriorism: wherein, by looking at the *exterior*, we *imagine* the interior; for the face often tattles on the heart, and an empty surface may bespeak an empty centre (though this is not true of alligator eggs). The Moon has a stony face, while the Earth's face is a slaphappy burlesque — screaming flocks of peacocks here, and cloudbursts there, and spriggy merriment everywhere. Such an exhibition is possible only if inside itself the Earth has a core whose nickel density enables the planet not only to sport a moon but also to hold on to tiny flighty molecules. For these bouncing, shimmying molecules are Earth's genius, and they are harder to keep than moons. Cloudland has a core of adamant.

On behalf of those who feel vacant and uninhabited, to whom nothing occurs, who look up day and night from chalky dust into unrefracted blackness, who watch their plush blueheaded neighbours yielding splashy gullies and snow devils and excitable vespiaries and backsliding pinnipeds and heady cauliflowers and turtle centuplets and rosy squirrelfish swarming through Rapture Reefs: on behalf of unprofitable individuals everywhere, is the Moon ordained to ever be a shabby waste of rubbled regolith? Could it never scrabble together a genius like the Earth's?

What about molecule trustees, like the Sun? The solar wind blasts a plasma of particles throughout the solar system; could not some of these accrue upon the Moon? For not *all* atoms are wiggle-away; xenon, for example, is heavy and slow. It would make a nicely noncombustible atmosphere, of glowing lavender hue, and would make sound possible, albeit slow, so everyone's voice would drop several octaves and everyone would sound like walruses. And xenon is an anesthetic, so inhabitants would be blithe and amenable to dentistry. But the wind that bringeth the elements taketh them away; the atmosphere on the Moon is thinner than the thinnest vacuum we can contrive.

Halos cannot be affixed to the head with pins and clips. Marañón forests, hosting spinetail birds and purple-backed sunbeams and grey-bellied comets and velvet-fronted euphonias and long-tailed weasels, cannot be administered from without. Glory cannot be administered from without. Glory will only coalesce on a body wherein throbs a fiery, molten, mad-stallion heart so dreadfully dense, so inescapably attractive, that it matters little the circumference of the frame.

Of course if your heart is *too* fervent, you will become an attractive incinerator, like the Sun, glorious but no pleasure boat. The glory of the Sun is violent and uninflected; its features are all flames and its sounds are all explosions. The Sun is so loud, like a million bombs all the time, that fine-spun sounds cannot be heard, like birds wading or figs tumbling or the muttering of mathematicians. On the Sun all private qualities disappear into the main loud yellowness.

Nothing makes a sound on the Moon and nothing ever could: not a harpsichordist, not a shattering tureen of mangel-wurzel stew, not the pebble-sized meteoroids that whang down at seventy-eight thousand miles an hour and heat the ground so hot it glows like a little piece of star; not the huge meteoroids that fracture the bedrock, forming craters two hundred miles across, creating new rings of mountains, making the Moon to tremble on and on — since it doesn't have a sturdy core the Moon is very convulsible; once atremble, it stays atremble. But it fractures and trembles and glows in absolute silence, for sound is like birds and cannot travel without air.

From looking at its face we had inferred that the Moon's heart is small and dead; but this is not to say that its face has no properties; not even the most stuporous face has no properties. The moonscape is pleated and rumpled, with rilles and ridges and craters and crevices and darknesses and brightnesses. Except for some meteor-made bruises, though, its features have not changed for three billion years; they are memorials of an ancient vim. Once the Moon was welling up from inside, jutting into volcanoes from the force of its own melting, cracking at the rind from its deep inner shifts. Now it wears the same glassy expression eon after eon, like a taxidermied antelope. The Moon is a never-brimming eye, a never-whistling teakettle; and it shadows the very flower of planets.

There are several kinds of orbits in the orbit catalog. One is an interrupted orbit, which describes the path of a dumpling flung from a window; the ground being the interrupter of the dumpling's orbit. Another is known as an open orbit, where an unaffiliated travelling object gets pulled to another body, curves around it and flies away never to return, like a minute. It is just a gravitational encounter and it merely redirects the object. The other kind of orbit is where a rock, after ages of streaking obliviously past acquisitive black holes and great gassy mooncatchers like Jupiter, happens to come close to a small motor-hearted globe, close enough to feel its influence, to be drawn closer, to make a circle around it, and another, and another, and never thereafter to stop, not for billions of years. Once it was its own and now it is a foundling. This wrapping of the one around the other is called a closed orbit.

In truth the beginning of the Moon is a secret: maybe a piece of Earth broke off and went into orbit; maybe the Moon was begotten by a terrible collision; or maybe it really was a drifter snatched from its onward way. However the Moon began, here is how the

Moon will finish: in a billion years the Earth will have nudged it far enough away that it will look 15 per cent smaller; in three billion years it will look smaller still; in five billion years, the Sun will become a red giant and swallow its children up. The Earth's involvement with the Moon will not last long enough to end.

The disposition of the universe – that crazy wheelwright – designates that we live on a wheel, with wheels for associates and wheels for luminaries, with days like wheels and years like wheels and shadows that wheel around us night and day; as if by turning and turning things could come round right. For the moment, if you are still in the field of feathery grass where you were playing the Moon, you might look back at your footprints. The Sun spins in place so his path is just a point; and the Earth leaves a long ellipse around the Sun; but your path is a convoluted zigzag, for you loop around a looping planet. Your trajectory is something like the trajectory of sea ducks. Little harlequin sea ducks swim over the oscillating waves of the sea, diving down into the cold grey-green waters to unfasten limpets and blue mussels from their rocks, swinging back up into the rough winter waves, the sea itself rolling up and down under the spell of the sailing Moon.

Twinkle Twinkle

. . . remembering that we are a part of universal nature,
and that we follow her order.

—BARUCH SPINOZA

For a while you were not allowed to go within four cubits of Spinoza, because of his incendiary ideas; getting close to him was like getting close to a fire. He had to abandon the dried fruit trade and find a more solitary job. Spinoza died in 1677, at forty-four, of a lung disease exacerbated by breathing the glassy dust generated in his work grinding telescope lenses. That was the same year Edmond Halley of Britain first made a note of Eta Carinae. It was an apparently ordinary star — not too bright, not too faint — of the fourth magnitude. Fourth magnitude is 2.5 times as bright as fifth magnitude, which is 2.5 times as bright as sixth magnitude, and so on: seventh magnitude is too dim for the untelescoped eye to see.

Many students of the stars begin with the Sun, that reliable yellow star outside your window that makes you hatch out of your sleep every morning and makes all your houseplants grow lopsided. But I would recommend starting, instead, with Eta Carinae, for it is erratic and could expire within the next million years.

By 1730 Eta Carinae had brightened to a magnitude of two, then faded back to four (1780), then brightened to two (1801), then faded to four (1811), then brightened to *one* (1827), then faded to two, then brightened to *zero*, then brightened *again* until it was shining at a magnitude of *negative one* in April of 1843. It outshone all the stars in the 1843 sky except for Sirius (and the Sun). Other stars twinkle-twinkle; from twinkle to twinkle they are the same. But if Eta Carinae is the little star of your song, then sing it like this:

'Twinkle twinkle TWINKLE *twinkle* twinkle twinkle. TWINKLE little star',

for the dynamics of your song should ideally correspond to the magnitude of your subject; of a temperamental twinkler sing a temperamental song.

If we had commenced our lesson with an everyday star, like the Sun, the story would have gone like this: five billion years ago a lump of gas and dust attracted more and more gas and dust until the gravity was tremendous. The gas and dust became so compressed that the atoms started fusing and giving off light and heat, qualifying the object as a star. The star shines for ten billion years, then changes into a red giant or a white dwarf. Since you, however, were born halfway through the part where the star shines for ten billion years, you assume that the star is incorruptible. You are a moth and the star is you. But with all its turmoil, its throbs and outbursts and spectacular fluctuations, Eta Carinae permits no such assumptions; you are a moth and the star is the Ingenious Gentleman Don Quixote.

If Eta Carinae were our Sun, it would engulf Mercury, Venus, Earth, and Earth's inimitable snowmen; it is a hundred times as big as the Sun and four million times as luminous. Smaller, more abstemious stars live to be ancient sturdy little chaps, while huge hellions burn themselves up in about a billion years, explode in supernovas, then fade away. Turn out the lights, the party's over.

After it brightened like a supernova, in 1843, Eta Carinae faded like a supernova, and from 1900 to 1941 it was down to the eighth magnitude. It seemed like everybody either was passed out on the floor or had gone home. But then the party resumed, and by now Eta Carinae is shining at fourth magnitude – or, to be more accurate, by the fifty-fifth century BC, since the star is seventy-five hundred light-years away. Living in a galaxy is like living in a neighbourhood where the house down the street might have burned down four thousand years ago but you wouldn't know it for another three thousand years.

A true supernova can be confirmed by the quietus which follows it. A star experiencing a real supernova does not revive in a hundred years; it has been atomised, returned to dust. Eta Carinae is a *supernova impostor*. They say that impostors are out to fool you, but many impostors are probably unaware of the expectations being excited and frustrated by their actions. Santa impostors who never squeeze down the chimneys, deposit the presents and eat the cookies probably just enjoy taking sleigh rides in the air, though the frivolous use of reindeer seems unconscionable.

In 1995 the Hubble telescope took a photograph of the mad subject of our study, which seems to be trying to outlive its end, to survive itself: a crimson glow of stardust surrounds two blue-white lobes — exploding lobes reflecting light, each much bigger than our solar system — with a thin disk of blue-violet lasers separating them; two cauliflowers end to end, girded with a translucent lilac tutu. This stupendous figure is thought to be formed by the paroxysms of Eta Carinae and a secondary star and the shocking collision of their respective winds. A wind-wind collision is similar to a sneeze-sneeze collision but more vehement. The incandescent cauliflower-ballerina is made of dust plus deep light; take away either ingredient and you have no celestial vegetables tripping the light fantastic in a laser tutu. Dust is light's finest medium and light is dust's most dithyrambic tenant, along with dwarf Roborovski hamsters.

Dust in space has to be deeply lit before we see it, which is different from dust on Earth. Earth dust will show up in the puniest lighting. Dust on Earth makes green less green and glass less glassy and thus subtracts quintessence. An angel covered in dust looks like a lump.

It is true, dust supplies clouds with essential rain-making material: raindrops will not form around nothing; they require a speck of something. But more dust does not mean more rain but tinier raindrops. Very dusty rain clouds bestow but bony raindrops and drift on, full of water. More dust means less rain, and less rain means more dust. The dust it dusteth every day. Too much dust makes birds fly away; if a bird stays behind, it will have to build its nest out of barbed wire. Too much dust means rabbit-stricken farms and trouble-stricken tomatoes. Particles of dust storms rub together, creating enough static electricity to kill the dewy, sleeping tomatoes and turn them black. Too much dust does the same thing to the dewy lungs, it takes the breath away.

Even in rainier areas, where dust is less inexorable

and submits to brooms and rags, it is generally detested, because dust is not organised and is therefore considered aesthetically bankrupt. Our light is not kind to faint diffuse spreading things. Our soft comfortable light flatters carefully organised, formally structured things like wedding cakes with their scrolls and overlapping flounces.

It takes the mortal storms of a star to transform dust into something incandescent. Our dust, shambling and subtractive as it is, would be radiant, if we were close enough to such a star, to that deep and dangerous light, and we would be ravished by the vision — emerald shreds veined in gold, diamond bursts fraught with deep-red flashes, aqua and violet and icy-green astral manifestations, splintery blinking harbour of light, dust as it can be, the quintessence of dust.

The Wild What

Several centuries ago the stars reconstellated into figures more relevant to the times. The Earth had been industrialising, mechanising, electrifying, while the stars were still trotting out swans and goats and bears every night. Men of the world advised the stars to update their subjects, to figure forth printing shops and electricity generators. Obligingly the stars complied, and for a while the sky was up to snuff; the stars were sophisticated and worldly; but then shops supplanted shops and machines surpassed machines and the sky was left behind, littered with musty antiques.

Thereafter were the stars persuaded to depict compasses and quadrants, stripped of their names, given numbers, all but regimented into a grid, before they had had enough and reverted to their old subjects:

dogs, dragons, herdsmen, bears. Take heed, worldly fashion – someone may trust you up to a point, but if you push him too far you will lose all the power you ever had over him and he will blaze up and turn into a bear.

The bear in the sky is sometimes mistaken for a ladle or a prawn, or the government, while the bear on the ground rarely is. There are a few discrepancies between the bear in the sky and the bear on the ground – for one thing, bears on the ground are not nocturnal; nor do they have long tails; nor are they stalked by ravening chickadees who cook and eat them once a year. (Chickadees are good cooks but they do not usually own cooking pots.)

The big starry bear is trailed by a little starry bear, about the size of an autumn cub padding around on plantigrade paws. Late autumn is when Earth bears and their children run out of fresh apples and honey, when they might come across a heap of fermented apples, and devour them, and lose their bearings. Bears on the ground are the most sweepable bears off their feet.

Bears in the sky are next. Many of the stars in the bear are leaving the bear: they belong to the Ursa Major Moving Group. If you saw an assortment of red berries in the air, all floating the same way and perfectly maintaining their configuration in relation to each other, you might surmise that they were all growing on the same invisible drifting hedge. Sometimes, in the pool, dispersed among the randomly paddling people, is a secret synchronised swimming team, not singing and smiling and exhibiting their legs but all heading the same way and all possessing an inward resemblance if not the same mass. As they move across the pool it sometimes looks like they are part of miscellaneous social clumps, but watch carefully and you will be able to discern that they are associated with each other and share a common drift, perhaps toward the slide.

That is what the Ursa Major Moving Group is like. Ostensibly members of the bear and the giraffe and the water carrier and the rabbit and the harvest maiden, these stars are secretly committed to the UM Moving Group. Like brother and sister berries, the stars of the UM Moving Group are chemically homogeneous, with unusually high levels of yttrium, and they came from the same cloud. They are slowly drifting toward Sagittarius; as they drift they will wrench apart the bear, the giraffe, the harvest maiden, the tresses of Queen Berenice, Apollo's goblet, the man in the coils of a snake, and the snake itself. Thus are many identities, over time, shown to be temporary alignments of components involved in a deeper allegiance. Goodbye to my goblet, goodbye to my bear; identity must yield to deeper identity. Goodbye to my giraffe, goodbye to my girl; local association gives way to an association of travelers across the firmament.

Stars, like thoughts, are not inevitable. Out of the diffuse disorder something may or may not coalesce, and floating specks in space find each other very escapable. Think how easy it is to escape the gravitational field of an animalcule. When consolidation does happen, it is usually precipitated by an outside force: a density wave, a nearby supernova, two colliding galaxies send the specks reeling, clustering, concentrating into collapsing factions, and those specks that once were strangers, easy come easy go, are now drafted into the same turbulent, raging-hot, high-pressure project — not just pressed close but pressed *into* each other, their previously repulsed protons fusing, four hydrogens becoming one helium. Out of these violent conjunctions are born the least violent, most oblivious things in the universe — neutrinos, rushing by the trillions through your person every second. Runners-up are oblivious to persons, tarantulas, silver and gold, landslides, dust bunnies, disapproval, hearsay, the cheese cart rolling by, but neutrinos are oblivious to all this and *daisies*.

The other by-product of nuclear fusion, besides neutrinos, is light. All bodies are radiant but not all radiance is visible: stars radiate visible light; planets and donkeys and couches radiate infrared waves. (If

your couch is emitting visible light GET UP IMME-
DIATELY!) Some condensing assemblies in space
never get big enough to radiate visible light; they just
condense. A star will not shine until it has assembled
enough self; once it has enough self it cannot help
but shine; once it starts to shine it cannot help but
burn the self up and blow the self away upon the stel-
lar winds. Some stars are so windy they lose a Sun's
amount of mass every hundred thousand years — at
that rate, if you weigh a hundred pounds, you could
be selfless in two yoctoyears.

Dubhe, a red giant on the bear's back, is not a
member of the Ursa Major Moving Group. In fact
it is drifting in the opposite direction. But Dubhe is
not all alone in the universe: Dubhe has a companion
star, Dubhe B. If your name is Ruby and you would
like to have a companion star, find someone named
Ruby B and place her in a strap and swing her round
and round. At first you will feel like you are doing all
of the work, but after a while Ruby B will start recip-
rocating and you Rubies will be a mutually-slinging
sensation.

Yes plus No equals a circle, where Yes is coming
together and No is flying apart. Two stars in mutual
orbit feel equally the forces of Yes and No, of gravity and inertia. If Yes were stronger they would crash
together; if No were stronger they would go tearing

off into the wild what. Ambivalence is an engine, a motion machine.

Three ambivalent planets have been detected circling 47 Ursae Majoris — a star between the bear's back paws — three messages dispatched thereto from Evpatoriia Planetary Radar. One of the messages is a musical programme created by teenagers and performed on the theremin, an instrument you play by waving your hands around in front of it. The First Theremin Concert for Aliens begins with 'Egress Alone I to the Ride' and concludes with 'Kalinka-Malinka'. Of course music does not always register with its intended audience — when music comes out of nowhere like that, it is hard for an audience to know whether they should admire it or not. The aliens might ignore the Concert for Aliens; then it will seem pointlessly conceived, spaceborne compositions transmitted in vain: the First Theremin Concert for Nobody, with a long way to go before it disintegrates.

But until something disintegrates, there is always a chance it will be taken to heart. Even if the aliens despise the theremin concert, perhaps on a leguminous exoplanet farther out, a bean farmer possessed of two muddy acres and one lost pig will be out late one night looking for his pig, calling 'Dewey, come back Dewey!' and the strains of 'Kalinka-Malinka' will reach him and he will look up astonished, into the wild what, whence cometh weird delights. The song is his who hears it.

The Bright Bear, made of stars and planetary nebulas and far-off galaxies, contains within it a Dark Bear, made of black holes and dust and gas and planets and moons and dwarfs. Black dwarfs are stars that have gone to seed: having run through all their hydrogen, all their helium, having stripped all the electrons off of their atoms, they are cool and spent and invisible. However at present black dwarfs are invisible not because their lights went out but because there are no black dwarfs. For any black dwarfs to exist yet, their precursors would have had to be older than the universe. Roomy an inn as it is, the universe turns away anyone older than itself, perhaps because it would alarm the other guests. Black dwarfs are as hypothetical as dark elves, except dark elves have no precursors.

Although brown dwarfs are dim and cool like black dwarfs, it is not because they are spent – you can only be spent if you once had something to spend. Brown dwarfs were always brown dwarfs; they start out dim and get dimmer; they are like pits that never had a peach. But even failed stars may have planets, though of dubitable habitability. What a brown dwarf can offer its satellites is stability – whereas hot young stars will explode, squandering their gravity, frizzling their followers, brown dwarfs are stable and attractive forever. But for all that, brown dwarfs are never adored; for a host star to be adored, it must first give ferns enough confidence to come out of the ground. The ferns on their planets never come out of the ground; the squirrels have no ferns to play under, and no squirrels to play with.

If there is seeing without perceiving, there is also perceiving without seeing. If Ruby B were invisible we could infer her presence from your anomalous wobbling. Brown dwarfs and super-Jupiters and black holes, though hidden from sight, can be inferred from the anomalies they cause in their seeable neighbours. Much divergent behaviour, in fact, is caused by invisible companions, although implied existence is not certain existence. Not everyone who stumbles stumbles upon an invisible bandicoot. Our Sun may not have a brown-dwarf companion called Nemesis which provokes our planet's periodic disasters.

It turns out that a lot of normal behaviour is also caused by the invisible, for example the fact that the galaxies do not fling their stars away like slingstones. There does not seem to be enough material to hold a galaxy together; they are rotating too fast – their outer stars should be thrown in various directions, like you and Ruby B if the strap broke. For galaxies to be as gravitationally stable as they are, spinning as fast as they do, they must contain a lot more matter than the matter we can see — about 80 per cent more. The galaxies have some pumpkins up their sleeves.

Is this true on smaller scales too? Apart from a visible fragment is everybody largely invisible — invisible like the magic part of magic mushrooms and the song part of songbirds? Maybe the balance between one's visibility and invisibility is like the balance between the salt and the water in the blood, delicate and critical, as becomes obvious when the balance deteriorates: people with an invisibility deficiency seem like paper dolls, subject to crumple. Other people have the opposite problem: they cannot be seen building a bicycle, nor making lentil soup, nor knitting a green wool sweater by candlelight; neither can you look down from your second-storey window in the morning and see them tromping off through the snow.

The Pinwheel Galaxy, a little east of Chickadee, is about twenty-two million light years away, so we see it as it was twenty-two million years ago: a swishing pool of lampy champagne, yellowy-pink, trailing fizzy streaks of sapphire stars. Space champagne is strong — Pinwheel champagne strong enough to have sparkled up a trillion stars. If we could be in the Pinwheel Galaxy now, peering back at Nebraska, that would also be a sight to see — the horselife and the hickorylife and the ducklife twenty-two million years ago, the horses tiny and amateur, with three toes on each foot, the

hickories following their own counsel, the ducks upended in the pond, dabbling for weeds, hoping never
to be ambushed by an ysengrinia.

The Pinwheel Galaxy seems to be dominated by
Euphoria. In fact it looks like most places out there
are dominated by one spirit alone, where none of the
others can get a foothold — see, for example, comets
commandeered by Confidence, or the intergalactic
stretches languishing under the rule of Patience. Not
even Mercury is really mercurial. But here on Earth,
Glee and Delinquency and Grimness share terrain.
Of course none of them has changed its essentially
rapacious nature; we still see how ruthless the spirits
can be, as when Joy possesses a dog and whacks her
tail against a wall, over and over, although the dog is
whining and her tail is broken. Many times a day a
mood sets up a monstrous dominion in the mind; but
before it kills us another takes its turn, and no ruler
who hands over the reins like that can be called absolute.

Maybe after thirteen billion years of experience,
even despots start to understand that despotism elimi-
nates anyone sensitive, that what they long for, more
than territory, is sensitive territory, that too much
electricity fries the wire that carries it, that in a wilderness of ice, or dust, or flame, there is no substrate for
Dread or Giddiness to grow on. If you were Reticence

would you rather inhabit a pseudostar or a waif, even if you had to share the waif with Warmth? So they check themselves, dwelling in combination, being as gentle as they can be, here, as gentle as they can. Translucent green worms hang down from tree branches, twisting on threads in the breeze. Male deer rub the velvet off their antlers. When the sun shines through the rain the drops turn clear gold.

Comfortless

My Aunt Stella's down comforter arrived from Texas, finally, but smelling scorched, leaking feathers. A little lightbulb in the car door burned a little hole in it. I have Band-Aided it up, but the feathers still escape because feathers want to fly. For fifty-three years Stella had been fastidious with the comforter, folding it away every morning in its plastic zippered case, when all it ever wanted was to get a little hole burnt in it. Goodbye seams, say the feathers: we are going to float!

When I am through losing my Aunt Stella's exultant feathers, I am going to fill the antique rose-coloured comforter with sand and double Band-Aid the hole. I will use clean, fine Jamaica sand. Sand will never float away and leave me bereft. Sand will be fun to sleep under.

Sand will find the hole, when the Band-Aids pull

away one night, and spill. I will wake up with hand-
fuls of sand spilling out of my hands. Goodbye seams,
goodbye hands, cries the sand: I am going to spill!

'Unkind hole!' I will cry. 'Obstinate comforter,
letting all my feathers out, letting all my sand out!
How can I depend on you? Must I ever be tending
to you, my difficult comforter, ever filling you with
something new, which only ends up floating away or
spilling? Must I live with no certainty, no continuity?
I will call you Comfortless!'

In my exasperation I will run to the store and buy
many chocolate mints and stuff my comforter with
mints. 'If you are going to have a hole,' I will tell it, as
I resolutely stuff it full, 'I will turn you into a chocolate mint dispenser. When you release your contents
(as you are always doing), I will get a treat.' It will be
a cunning plan, inspired by sheep, who also eat their
bedding. But it will not be quite cunning enough:
chocolate mints, unlike grass, melt, and do not hesitate to coat a warm sleeping person with goo.

I will wake up as Chocolate-mint Person, I will
stumble to the door, unhappily attracting sand and
feathers on the way; I will stand on the lawn; I will look
up at the stars and bleat, 'Stars! I am having trouble
with my comforter! You are so serene! How can I be
serene like you?' They will look at each other know-
ingly, for they have answered this question millions of

times. And then they will twinkle back to me, 'Person, you will never be like a star. Things for you will always float away and spill and melt. The closest thing to serenity, for you, is laughing.' I will recognise this as true. I will stand there, just another sandy, feathery, chocolate-mint person laughing on the lawn.

THE
ORACLE

It used to be, if we had an important question about life we could visit the oracle, perched on her tripod over a vent in the earth. We would bring her honeycakes of devotion and ask her: 'Should I cross the river tonight?' 'Why is there a pall over my heart?' or 'Should I marry Cyril?' She would wait for the earth to exhale and then, exhilarated by the vapours, she would reply, 'Oaths are like paper feet' and 'Find where the goat loves saltwater' and 'Wing-a-willow-way-away'. Her messages were as intelligible as the jingly messages of wind chimes.

But now when we visit the tripods all we find perched there are cameras, hardly reminiscent of oracles — they are literal thinkers, like most anyone we ask for advice. 'You should cross the river tomorrow because all the ferries are full tonight.' 'You feel depressed because your riboflavin intake has been in-adequate.' 'Cyril is a good choice for a husband. He is involved in the community and has never been con-victed of a felony.' Such concrete answers; sometimes one misses the old inspired answers. Why should con-versation always be so much more coherent than ex-perience?

Perhaps you have noticed, when you take your wind chimes down to polish them, that the wind does not stop blowing, or that when you put your flute away you do not stop breathing. The wind does not need wind chimes to blow, nor does a person require a flute to breathe; the oracles were not speaking from their own understanding but transmitting the Earth's emanations. They were mediums, exhilarated inter-mediaries – the middlewomen gone, the Earth itself may be our authority, what communications we re-ceive from it as cryptic and ravishing as the ravings of Pythia: a frog or a fox flying by, Texas mud babies in the bog, Chinese lantern plants, chrome yellow foam resembling scrambled eggs but itinerant and not good with toast. Who needs a priestess with the divinity at hand?

Especially oracular are mushrooms, defiers of culture and transcenders of understanding. If ever you grow weary of concrete, so much concrete con-versation, you might take your questions to the forest. Pale brown mushroom snouts poke through the dirt; whiskers with red pinheads emerge on rotting logs; false puffballs, true puffballs appear, spongy white podiums for miniature lecturers, red rubber eggs throwing off their universal veils, orange peels with eyelashes, orange-yellow jelly babies, green glowing mushrooms, dead man's fingers, witches' butter, shaggy manes, self-liquefying mushrooms, mushroom chaos.

Mushroom chaos is not always visible; among the many emanations of the Earth, some do not come to the surface. Truffles have an underground habit: to discern them you need a truffle puppy and a hazel grove. Other mushrooms are visible with invisible consequences, like gold caps which induce spirituality, jack-o'-lanterns which induce stern pain, destroying angels which induce the afterlife, red-capped mushrooms with yellow warts which induce meaninglessness in insects; or turkey tails, pretty pink-and-green ruffles growing out of trees, decomposing their hearts.

As they grow tall, some trees abandon their lower branches, to concentrate on more tallness, absurd tallness. The rejected branches below break off, leaving gashes behind. In the winter, starving deer tear branches off of trees for meals of sticks, and snow and ice can sever a limb, being too heavy for it to bear.

A tree can be tempted out of its winter dormancy by a few hours of southerly sun — the readiness to believe in spring is stronger than sleep or sanity. The warmed section swells and grows green under the bark; then goeth sun under hill and the foolish green freezes, cracks. Turkey tails feed on scars like these, decomposing them, with beetle crews arriving expeditiously to excavate the area. Once the tree was full of tree; now the tree is full of holes. It is not good to pour concrete in the holes. Would you pour concrete into your own wounds?

Pelicans whose duty it is to glide, prairie grass whose duty it is to bend, spiderlings whose duty it is to be carried on their gossamer strands into the upper atmosphere — are as familiar with the wind as if it were their wife. But for the tree, whose duty it was to stand, the wind was only a passerby, like Mr. Umpleby and dog on a ramble through the woods. Then misfortune and mushrooms came along and punched holes at intervals up and down its trunk, preparing the way for the wind, and elements long kept external were enabled to enter in. Now the tree is full of air, and bear, and meadow mouse. Thus does a solid tree become a channel, a medium, an instrument.

AFTERWORD

The Round-Earth Affair

We were each like a tree grown in a cage. My neighbours and I slept tucked behind thin gates and thin doors, beneath ceilings, stacked and walled away from nature, all light from moons and planets undetectable through the smog. Up until a year ago, what we met of nature during the day was kept in pots or disguised as breakfast. Most of the time we felt we did not need nature; some of the time, when it thundered or trembled, we felt that we did not *want* nature. And some of the time we felt that we did not know what we needed or wanted.

We had heard that bad things were in store for nature, the way the oceans are overfished and the skies fouled, but in my city people are not so horrifiable: we'd seen movies of the future, when the Earth is finished being ruined — humans wheezing out some last

163

haggard weeks in cement tenements — and yet to us these conditions didn't look so strange. We were hard to horrify, living already as if nature were gone.

'The round-Earth period can only last so long, anyhow,' we'd say. 'The Earth used to be flat, and for now the Earth is round, but the Earth is going to be flat again — faded and flat as a bell that thuds and does not ring.'

And so it was useless for anyone to try and sign us on with the ranks of the environmentally panicked, or the environmentally conscientious, or even the environmentally uneasy. We were resigned to losing Planet Earth, whose silence, whose spaciousness, whose Newfoundland, whose whales, were as foreign to us as Planet Huffenpuff's over in the next galaxy. Imaginary, or very far away, was how we saw 'Earth', the thought of whose demise brought some folks to weepy shrieks.

But while resigned people can be infuriatingly unresponsive to shrieks, they are, in some ways, very easy to enchant: when you expect to lunch on roaches and granite, bread delights you.

There was an infirm building across from my apartment that the city finally, two years ago, smashed up and carried away, leaving a pitiful square of trash. Then, a year later, they started a park there. The park was pitiful too at first, and since the plants had to endure that winter as infants, my neighbours and I stayed unacquainted with nature for some months. But in the spring the tender and undefeated thin green fingers emerged from the ground and from the stick-trees, and we began to notice the park — even if we were just hurrying by — because bits of it were straying into our paths.

In May, some small lilac butterflies started to break up the tedium of road. On my way to catch the bus, road, road, road, on my way back home, road, road, road, on my way to catch the bus, road, road, road, butterfly. A *butterfly*, a giddy looping thing which appeared to be out of its mind. It flew very energetically but made little progress, flying in weird and impractical patterns. It was not gently, serenely, carefully floating through the air, like on greeting cards. This butterfly seemed magnificently crazed. It was as if a creature crawling around on short legs had suddenly obtained light purple wings and somebody had forgotten to tell it about the training.

One morning I saw a tiny frazzled bird running back and forth by the side of the road, tormented by the kerb. It must have been prematurely unnested — it couldn't fly, couldn't jump, could only dash, and so all its worried energy was manifested horizontally. It had fallen off the kerb and had no way to get back over it. I stopped and picked it up — small, bony, and soft — and left it dashing around in the grass. Whether people need nature or not, it was clear that *nature* needed *people*: I'd seen a man picking a straying green-brown beetle up from the road and putting it back in its park, and I had seen beetles that didn't get picked up from the road: his beetle, and my bird, absolutely needed people.

But perhaps nature needs us like a hostage needs her captors: nature needs us not to annihilate her, not to run her over, not to cover her with cement, not to chop her down. We can hardly admire ourselves, then, when we stop to accommodate nature's needs: we are dubious heroes who create a peril and then save its victims, we who rescue the animals and the trees from ourselves.

People need nature, at times, in the same way: since fire and snakes and storms can so easily make bones of us, we need them to be merciful. But if destructive potential were the only reason that humans and nature needed each other, each would have a reason to be polite, but not much reason to be miserable if the other vanished. Nature, however, seems to inspire in us something more than politeness.

On Sunday mornings, for a long time, I have taken my chairbound neighbour Demetrio for walks. Before the park came, we took a look at people visiting the newsstand, people visiting each other, people buying bread. Now that flowers bloom across the street we look at tall red billowy flowers and Demetrio looks astonished, like he is in love, like an old man in love with a flower.

There is a print in our library of a painting by Caspar David Friedrich, called 'Monk by the Sea'. Most of the painting is sky, a north brilliant ice-blue sky. A small figure stands on ice, looking out across rough black ocean. No boats are heading off, and no boats land. It is not a sea for sailing. It could swell up and suck the man down; the sky could whip around and send him slipping into the water; the ice could crack and float away with him. The scene he inhabits has enough power to terrify and extinguish tens of thousands of monks; yet the painting is not of 'Monk Running Away from the Sea', and while he *is* painted with his back to the observer, it is conceivable that he is *not* standing there screaming. Nature has another power besides the power to terrify.

Because it is so empty of human commotion, the painting of the monk used to seem archaic, if not incomprehensible; but even though the proportions are reversed in our neighbourhood, and nature is now the vulnerable, quiet one surrounded by an immense ocean of persons, the sense of the painting has been resurrected for us; for when she is planted across the street, we are fastened with desire: we cannot relinquish butterflies and return to uninterrupted road. We have lost our composure, for the Earth is still round, and the Earth is still ringing.

GLOSSARY OF
STRANGE BEASTS
AND
PHENOMENA

Glossary

MOULDYWARP Small sightless animal; dwells complacently in the dirt; also goes by Mole. May live near mummichogs.

FEAST OF THE BEAN-KING Midwinter feast where everybody gets as fat and featherbrained as possible in one night and one lucky devil finds a bean in his slice of cake.

SAGITTARIES Centaurs – half-man and half-horse – who specialise in archery. Like everyone else, centaurs also specialise in their own smell.

VASTY (As differentiated from 'vast') Has approximately the same meaning as 'biggy', 'hugey', and 'giganticky'. Do not let anyone tell you these words are not words; all words are words.

TOXODONS If ever anything seemed indestructibly built, it was the toxodon, chunky and solid and stout. However tumbleweeds have fared far better. Something to think about, trucks.

FLOPPY KID SYNDROME; MAD STAGGERS Some goats live long enough to lose their hearts to moss and peas and sparkling sleet and some goats do not.

THE QUAGGA Browsing animal that resembled a zebra, except with slipperier stripes that fell off its round brown bottom. Quaggas themselves were slippery and fell off the face of the Earth.

MUSPELHEIM; NIFLHEIM You know how when you hold an ice cube over a candle flame, a petunia appears: well, the commingling of Muspelheimers and Niflheimers was a cosmological demonstration of this simple combinative effect. As retrospectively prophesied by Snorri Sturluson.

DRAGON-GAGGERS Instead of swords, some heroes wield toothbrushes, say they are there to 'brush the dragon's teeth', and then they poke the toothbrush into the back gaggy part of its hot tongue.

SMITHERIES AND SMELTERIES First the smelter smelts the iron from the ore, then sends it over to the smithy who smiths it into an automobile. After that the automobile is sent to the waxery where it is waxed, and from there it goes to the rustery where it rusts.

CROCODILOPOLIS Egyptian city near Memphis where they worshiped that jewel of animals, the crocodile. The sacred crocodile Petsuchos had his own temple, swimming pool, and sandy beach and he got a nice new body every seventy years or so.

THE DAY OF THE LILY PAD Ever since we hatched, our souls longed for this glorious appearance. Though our visions of it were hazy, we felt, we *knew* that we were not made for waterspouts, waves, and billows. And when the lily pads started to unroll, we said, Aha, aha.

ARGLE-BARGLE What, are professional bruisers like myrmidons going to *argue* with flowering vines, exchanging views, citing evidence, justifying positions? I don't think so.

RADISH MINISTERS Big muckety-mucks, radishes employ only the finest retainers to dance attendance on them. Radish Ministry is a plummy yet demanding office, quite difficult to secure, best inherited.

IMPENDING DOOM Some dooms tug at your sleeve, some dooms shimmer by, and some dooms bring you to your knees.

DR. OSIP IWASYKIW Definitely someone with the pedagogic itch. You have to be careful what you say around pedagogues unless you want everything from dog brains to Neanderthal religion explained to you.

WHIMWHAMS Doodads; objects with no utility, no purpose; ceramic frogs, zebra figurines, crystal pea-fowl trinket boxes; frogs, zebras, peafowl.

THE AGE OF EFFICIENCY; THE AGE OF SANITY; THE AGE OF REFINEMENT Our inventions have long been ahead of us in terms of efficiency and sanity, productivity and predictability. Oh, how we've wished we could be manmade, too. What has been keeping us back, keeping us messy? The animal impediment, within and without. Eliminating these impediments, we will surely be catching up with our machines, re-sembling them more and more impeccably.

SANDY BONES Mr. Sandy Bones has had more than a minor disaster. Sand is not supposed to have any concourse with bones. Something, probably unfettered singing, has dissolved that sanctuary that was his skin.

LUNCHEON LADIES As no samurai should go out without his topknot, no lady should go out without her zoo pass. Who knows when she might be invited to a spontaneous luncheon at the zoo café to celebrate the acquisition of a Malaysian babirusa.

ZOOLOGICAL PARK OF GHOULS In Loir-et-Cher, in France. Boring. Prohibited from grave-robbing, ghouls are just drippy.

FIREFLAKES The stars; as transitory as snowflakes only their transitoriness is protracted.

CLOUDLAND Earth, land of imagination.

MOLECULE TRUSTEES The sun and all of us are molecule trustees, administering the molecules entrusted to us until they are passed on. Like any trustee we do not own the property, nor do we decide who will receive what we stewarded. It might be somebody grumpy like Xanthippe.

MOONCATCHER It's not just moons — Jupiter is also up for catching ducks, death squads, anything flying by.

THE INGENIOUS GENTLEMAN DON QUIXOTE Don Quixote read too much and went mad, for literature magnifies the madness that is humanity. If you read too much and exacerbate your humanity even a moth will be able to discern your mutability.

RECONSTELLATE What stars and people generally do in response to the latest findings: rearrange themselves. Hedgehogs are examples of nonreconstellators, though, standing on their own four feet in the face of the headiest trends.

LADLE The long tail of the Great Bear is also the handle of the Big Dipper, which is an asterism: less distinguished than a constellation, lower down in the hierarchy of starry patterns. Any goose can make up an asterism. Constellations are superior to asterisms and asterisms are superior to asterisks.

THE WILD WHAT Anything unknown, for example an unborn baby. One day she may be a wild spelunker or a wild Lutheran but for now she is a wild what.

LEGUMINOUS EXOPLANET *Leguminous* means, basically, 'beany'. If we are going to posit alien beings in outer space, then please, let us posit alien beans for them to plant and eat.

PLANET HUFFENPUFF I don't know if it has whales or dark-green groves or waves beating down. But if there is a meadow on that world, with grasses that turn pale and curly when they die, there must be someone, even if it's just a cricket, who loves it.

Acknowledgements

Thank you to Karl Wilcox, Susan Lohafer, and David Hamilton for invigorating instruction. Thank you to Kerry Reilly, Mia Nussbaum, Erica Bleeg, Kirsten Giebutowski, Mara Naselli, Heal McKnight, Maggie McKnight, Alicia Holmes, and Ginny Wiehardt for illuminating conversations about reading and writing; to Brigid Hughes for imaginative assignments; and to Eula Biss and Sean Hopkinson for kindness and wisdom in reading these essays along the way.

Thanks to the Rona Jaffe Foundation and the Whiting Foundation for their generous gifts. Thanks to Blue Mountain Center and the Jentel Foundation for summertime in the forest and the hills.

I am grateful to Jin Auh and Luke Ingram for taking splendid care of my essays, and to Patrick Thomas and Jenny Lord for bringing this book into the world. Thanks to Mike Niebauer, Tim Bennett, Jay Bennett, Rick Franklin, and Dave Thomas for setting my words to music, and to Nate Christopherson for bringing my words to life.

Although they will never read this or any other book, I want to offer thanks to and for my fellow creatures, especially Annabelle and Beanstan, who teach me what it is to be ardently alive.

Thank you to Carrie Fry, Jody Hechtman, and Aviya Kushner for long and lavish friendship, and to Yiyun Li for showing me how to jump into deep water.

With all my heart I am grateful to my parents, Benjie and Sharon, for their inexhaustible love; thanks to my brother Bj for good ideas about the Great Bear; to Deb, Jonathan and Judy, Lori and Mark, Jennifer, Audrey, Lorna and Rick, and my dear grandparents, Jack and Frances Hardwick and Helen Leach — thanks to all my family, for their tremendous capacity for delight.

And thank you, Matthew, for taking me to heart.

Earlier versions of these essays appeared in the
following journals and books:

Orion Magazine — 'Radical Bears in the Forest Delicious',
    'The Oracle', and 'Pea Madness'
Tin House — 'The Wild What'
The Massachusetts Review — 'God' and 'The Safari'
Pushcart Prize XXXV — 'God'
Ecotone — 'Memorandum to The Animals', and 'Twinkle
    Twinkle'
The Los Angeles Review — 'Love'
A Public Space — 'Donkey Derby', 'Sail On, My Little
    Honeybee', 'The Round-Earth Affair', 'You Are Going to
    Fly', and 'Please Do Not Yell at the Sea Cucumber'
Best American Essays 2009 — 'Sail On, My Little Honeybee'
Gettysburg Review — 'Goats and Bygone Goats', 'Trooping
    with Trouble', and 'Talent'
The Not-Book — 'The Same Old Joy'
The Iowa Review — 'In Which the River Makes Off
    with Three Stationary Characters', 'Stairs',
    'Comfortless', 'When Trees Dream of Being Trees',
    and 'Trappists'
Identity Theory — 'Warbler Delight'

Typeset in Mrs Eaves, a contemporary rendition of the classic Baskerville typeface by English printer and punchcutter John Baskerville.

Interior design and typesetting by Gretchen Achilles/Wavetrap Design.